Handbook of Artificial Intelligence for Smart City Development

This handbook explores smart cities of the future, provides an understanding of their development and management systems, and discusses the technology challenges. It also discusses why and how humankind can benefit from them.

Handbook of Artificial Intelligence for Smart City Development: Management Systems and Technology Challenges covers the whole journey of the development of smart cities, from the foundations to the usage of technologies, including the challenges and issues that policymakers and technologists may face during the concept, development, management, and implementation stage. The book also covers sustainable strategies and safety measures and offers real-life cases of advancements in manufacturing approaches for smart cities. The book includes upcoming AI technologies such as big data analytics, block chain, machine learning, fault diagnostics, and a lot more.

This handbook is intended to appeal to readers from industry and research. The book is also meant for academicians and students from across many disciplines. Moreover, this book is a medium for consultants, government agencies, and policymakers to grapple with topics and perspectives outside their lane.

Big Data for Industry 4.0: Challenges and Applications

Series Editors: Sandhya Makkar, K. Martin Sagayam, and Rohail Hassan

Industry 4.0 or the Fourth Industrial Revolution refers to interconnectivity, automation, and real-time data exchange between machines and processes. There is a tremendous growth in big data from Internet of Things (IoT) and information services, which drives the industry to develop new models and distributed tools to handle big data. Cutting-edge digital technologies are being harnessed to optimize and automate production, including upstream supply chain processes, warehouse management systems, automated guided vehicles, and drones. The ultimate goal of Industry 4.0 is to drive manufacturing or services in a progressive way to be faster, effective, and efficient, which can only be achieved by embedding modern day technology in machines, components, and parts that will transmit real-time data to networked IT systems. These, in turn, apply advanced soft computing paradigms such as machine learning algorithms to run the process automatically without any manual operations.

The new book series will provide readers with an overview of the state-of-the-art in the field of Industry 4.0 and related research advancements. The respective books will identify and discuss new dimensions of both risk factors and success factors, along with performance metrics that can be employed in future research work. The series will also discuss a number of real-time issues, problems, and applications with corresponding solutions and suggestions—sharing new theoretical findings, tools, and techniques for Industry 4.0, and covering both theoretical and application-oriented approaches. The book series will offer a valuable asset for newcomers to the field and practicing professionals alike. The focus is to collate the recent advances in the field, so that undergraduate and postgraduate students, researchers, academicians, and industry people can easily understand the implications and applications of the field.

Big Data for Entrepreneurship and Sustainable Development
Edited by Mohammed el Amine Abdelli, Wissem Ajili Ben Youssef, Ugur Ozgoker, and Imen Ben Slimene

Entrepreneurship and Big Data
The Digital Revolution
Edited by Meghna Chhabra, Rohail Hassan, and Amjad Shamim

Microgrids
Design, Challenges, and Prospects
Edited by Ghous Bakhsh, Biswa Ranjan Acharya, Ranjit Singh Sarban Singh, and Fatma Newagy

Handbook of Artificial Intelligence for Smart City Development
Management Systems and Technology Challenges
Edited by Sandhya Makkar, Gobinath Ravindran, Ripon Kumar Chakrabortty, Arindam Pal

For more information on this series, please visit: https://www.routledge.com/Big-Data-for-Industry-4.0-Challenges-and-Applications/book-series/CRCBDICA

Handbook of Artificial Intelligence for Smart City Development

Management Systems and Technology Challenges

Edited by
Sandhya Makkar, Gobinath Ravindran,
Ripon Kumar Chakrabortty, and Arindam Pal

CRC Press
Taylor & Francis Group
Boca Raton London New York

CRC Press is an imprint of the
Taylor & Francis Group, an **informa** business

Designed cover image: Shutterstock

First edition published 2024
by CRC Press
6000 Broken Sound Parkway NW, Suite 300, Boca Raton, FL 33487-2742

and by CRC Press
4 Park Square, Milton Park, Abingdon, Oxon, OX14 4RN

CRC Press is an imprint of Taylor & Francis Group, LLC

Library of Congress Cataloging-in-Publication Data

Names: Makkar, Sandhya, editor. | Ravindran, Gobinath, editor. | Chakrabortty, Ripon K. (Ripon Kumar), editor. | Pal, Arindam, editor.
Title: Handbook of artificial intelligence for smart city development : management systems and technology challenges / edited by Sandhya Makkar, Gobinath Ravindran, Ripon Kumar Chakrabortty, Arindam Pal.
Description: First edition. | Boca Raton : CRC Press, 2024. | Includes bibliographical references and index.
Identifiers: LCCN 2023009098 (print) | LCCN 2023009099 (ebook) | ISBN 9781032125923 (hardback) | ISBN 9781032125947 (paperback) | ISBN 9781003225317 (ebook)
Subjects: LCSH: Smart cities. | Artificial intelligence.
Classification: LCC TD159.4 .H3634 2024 (print) | LCC TD159.4 (ebook) | DDC 307.1/416028563--dc23/eng/20230517
LC record available at https://lccn.loc.gov/2023009098
LC ebook record available at https://lccn.loc.gov/2023009099

ISBN: 978-1-032-12592-3 (hbk)
ISBN: 978-1-032-12594-7 (pbk)
ISBN: 978-1-003-22531-7 (ebk)

DOI: 10.1201/9781003225317

Typeset in Times
by KnowledgeWorks Global Ltd.

Dear Miraya

My loving daughter, you are an incredible child. You are my inspiration, and I am here because of you. I dedicate this book to you.

For my parents Mr. Sat Paul Bajaj and Mrs. Pramila Bajaj.

Words can hardly describe my thanks and appreciation to you. You have been my source of inspiration, support, and guidance. You have taught me to be unique, determined, to believe in myself, and to always persevere. I am truly thankful and honored to have you as my parents.

Sandhya Makkar

I wish to convey my heartfelt thanks to my wife Rajanimol PR and to my kids Natasha RG Nair, and Suryadev RG Nair for being with me always. Also, I wish to convey my thanks to my parents T. Ravindran and M. Gowri who are instrumental in my life.

Gobinath Ravindran

The book is dedicated to my parents and lovely family.

Ripon Chakrabortty

I dedicate this book to my father Bani Prasad Pal and mother Bani Pal for their love, care, patience, and support in bringing me up.

Arindam Pal

Contents

List of Figures and Tables

Editors

Dr. Sandhya Makkar is as a senior lecturer of Data Analytics in the Department of Digital Analytics and Technologies at Sheffield Hallam University, United Kingdom. Before that, she worked as the area head for the Department of Operations and Analytics at Lal Bahadur Shastri Institute of Management, Delhi, India. She has more than a decade and half years of academic and research experience in optimization, machine learning, and big data analytics. She secured first class throughout her academic career. She has authored and coauthored numerous research articles that are published in international journals, books, and conference proceedings. She has also been appointed as a book series editor with CRC Press, Taylor & Francis Group. She is involved in innovations and patent development. She can be contacted at bajajsandhya @gmail.com or visit the following website for more details: https://sites.google.com/view/sandhyamakkar/home

Dr. Gobinath Ravindran is presently serving as professor and head of the Department, Civil Engineering of SR University, Telangana, India and also a member of Center for Construction Materials and Methods division. He completed his PhD in the field of Environmental Geotechnology and Disaster Management from the Center for Disaster Management and Mitigation, VIT, Vellore, his ME in Environmental Management from the College of Engineering, Guindy, Anna University, and his BE in Civil Engineering from Bharathiar University. He has handled several undergraduate courses in civil and environmental engineering, including project management, smart materials and structures, geosynthetics, engineering mechanics and is presently handling elective courses related to disaster management and intellectual property rights in various universities. His research interests include sustainable materials, construction materials, landslide mitigation and management, soil bioengineering, engineering education, geotechnology, environmental geotechnology, sustainable development, sustainable construction technologies; he has handled several projects related to these domains on various levels. He has executed a good number of consultancy projects related geotechnical and environmental engineering domains. His research profile is available at www.researchid.co/gobinathdpi

Dr. Arindam Pal is a senior research Scientist at Data61 in the Commonwealth Scientific and Industrial Research Organisation (CSIRO), and a conjoint senior lecturer in the School of Computer Science and Engineering at UNSW, Sydney. His research interests are in algorithms, artificial intelligence, data science, optimization, and machine learning. He works on business and research problems at CSIRO and collaborates with faculty members of universities, both in Australia and abroad.

He earned his PhD in Computer Science from the Indian Institute of Technology Delhi. He has over 15 years of research experience in software companies like Microsoft, Yahoo!, and Novell. He has published academic papers in reputed conferences and journals and was granted patents in India, the USA, and Europe. He is a technical program member for several reputed conferences and a technical reviewer for many renowned journals. He is a leader in the area of algorithms, data science, optimization, and machine learning. He is a senior member of both ACM and IEEE.

Dr. Ripon K. Chakrabortty (Senior Member, IEEE) is an assistant professor in Systems Engineering and Decision Analytics at the School of Engineering and Information Technology, University of New South Wales (UNSW), Canberra, Australia. He is also the team leader and founder of The Decision Support and Analytics Research Group at the School of Engineering and Information Technology, UNW. He serves as the program coordinator for the Master of Decision Analytics and Master of Engineering at UNSW in Canberra. He has written two book chapters and over 170 technical journals and conference papers in multiple prestigious venues. His research interests cover a wide range of topics in operations research, project management, supply chain management, artificial intelligence, and information systems management. Many organizations, such as the Department of Defence and the Commonwealth Government of Australia, have funded his research programs.

Contributors

Pooja Ahuja
SVN University
Sagar, India

Jacob Alexander
Resources 4 Solutions Inc.
Delta, British Columbia, Canada

Suman Ananthanarayan
University of Mumbai
Mumbai, India

Debarati Basak
Carpediem EdPsych Consultancy LLP
Bengaluru, India

Anudeep-Sri Bathina
University of Massachusetts
Dartmouth, Massachusetts
and
Nikita Kalloorthottiyil Chacko (RECE)
Dufferin Corporate Centre
London, Ontario, Canada

Varun Chhabra
Quantum Age Solutions Ltd.
Middlesex, United Kingdom

T. Kalavathi Devi
Kongu Engineering College
Erode, India

Amol S. Dhaigude
Manipal Academy of Higher Education
Manipal, India

Riju Ghosh
Amity Business School
Noida, India

Mukta Goyal
Guru Nanak Dev Institute of Technology
Delhi, India

Ekata Gupta
Guru Nanak Dev Institute of Technology
Delhi, India

Manju Jose
Majan University College
Muscat, Oman

S.K. Kabilesh
Jai Shriram Engineering College
Tirupur, India

Soni Lanka Karri
University Brunei Darussalam
Bandar Seri Begawan, Brunei

R. Karthikeyan
Varadhaman College of Engineering
Hyderabad, India

Arshia Kaul
Carpediem EdPsych Consultancy LLP
Bengaluru, India

Vasundhara Kaul
Carpediem EdPsych Consultancy LLP
Bengaluru, India

V. Kavitha
Hindusthan College of Arts and Science
Coimbatore, India

Prabha Kiran
Westminster International University
Tashkent, Uzbekistan

Sinimole K.R.
Rajagiri Business School
Kochi, India

Vijayalakshmi A. Lepakshi
Reva University
Bengaluru, India

Manasvi Logani
Indraprastha College for Women, University of Delhi
New Delhi, India

Sandhya Makkar
Sheffield Hallam University
Sheffield, United Kingdom

Seyed Jalaleddin Mousavirad
University of Beira Interior
Covilhã, Portugal

Mooon Paiithannkar
Sinhgad Technical Education Society
Pune, India

Roli Raghuvanshi
University of Delhi
New Delhi, India

Rashmi Rai
Christ University
Bangalore, Karnataka, India

Prashanth C. Ranga
Conestoga Institute of Technology
Kitchener, Ontario, Canada

Gobinath Ravindran
SR University
Hyderabad, India

Ravinder Rena
Durban University of Technology
Durban, South Africa

A. Stephen Sagayaraj
Bannari Amman Institute of Technology
Erode, India

Tanushree Sharma
Manipal University
Jaipur, India

Abhishek Srivastava
Gopal Narayan Singh University
Jamuhar, India

B. Sundaravadivazhagan
University of Technology and Applied Science
AI Mussanah, Oman

Pankaj Tiwari
SVN University
Sagar, India

Prakash Kumar Udupi
Middle East College
Muscat, Oman

Asad Ullah
Middle East College
Muscat, Oman

M. Veeramani
Hindusthan College of Arts and Science
Coimbatore, India

1 Machine Learning for Smart City AI Systems

Manasvi Logani[1] *and Sandhya Makkar*[2]
[1]Plaksha University, Mohali, Punjab, India
[2]Sheffield Hallam University, Sheffield, United Kingdom

1.1 INTRODUCTION TO SMART CITIES

While stating the obvious things, smart cities comprise two words, "smart" and "cities". Thinking of the word smart, the first thing that comes to mind is a competent, self-sustaining and mature environment or person. Combining all these qualities with a technological perspective and the concept of cities is what is meant by the term smart cities.

More formally, smart cities are those that integrate the use of information and communication technology, also known as ICT, to improve the overall quality of life and optimize infrastructure sustainably while promoting economic growth.

These cities consist of a well-planned urban environment to create sustainable living for all. Pollution control, parking and traffic management, public security, energy conservation, healthcare etc. are some of the focuses of a smart city.

1.2 IMPORTANCE OF SMART CITIES

As the world is exponentially progressing towards technological advancement, it makes sense to leverage this for economic growth and better lifestyles. With this aim, there are numerous significances of smart cities. A few of these points are as follows:

1. Improved Healthcare Systems
   With an increase in automation, healthcare will become easily accessible with a more accurate prediction of diseases. With the use of previous data and certain tools, the classification of diseases will be easier and they might get diagnosed at an early stage, leading to be lifesaving. Also, doctor consultation will become more widely accessible through video conferencing appointments.
2. Better Safety Measures
   One of the main advantages of a smart city will be an extensive installation of CCTV cameras to collect data for incident detection, gunshot detection, license plate recognition, face recognition etc. All these facilities will reduce the crime rate considerably because the chances of getting caught will immensely get more than before.
3. Environmentally Friendly
   When it comes to human development, one of the most neglected aspects is the environment. Over the years, there has been a huge depletion of resources, air and water quality index etc. The main aim of a smart city is not just

DOI: 10.1201/9781003225317-1

development but also sustainable development. Therefore, to achieve this, air and water quality detectors will be installed to keep the pollution level in check, energy conservation techniques will be adopted, and a thorough check on resources such as electricity and water will be maintained to limit wastage.

4. Organized Traffic Systems
   With the advent of the advanced age, commuting from one place to another has become easier. The term "easier" here refers to in terms of the availability and accessibility of modes of transportation and the technology used. This is because, ironically, automobiles were created to reduce travel time but instead due to high traffic congestion, it has become difficult to cut travel time. Smart cities will focus on reducing this traffic to increase comfort, along with other facilities such as smart streetlights and traffic rules detectors.
5. Improved Infrastructure
   As the cities are growing older, their infrastructures consisting of buildings, roads, bridges etc. are getting weaker. In the smart cities, there will be an installation of certain sensors to collect data and report if there are any cracks or considerable damage to the infrastructure. With the help of this, it will become easier to prioritize development and even save lives in certain situations.

1.3 BACKGROUND OF STUDY

ICT is a discipline integrated with unified communications, telecommunications and computers for creating an infrastructure that enables modern computing. It includes radios, televisions, copiers, fax machines, telephones, satellites, multimedia, and internet which are all communication mediums in some or the other way.

Since the advent of ICT, there has been a gradual change in how the world is perceived. Initially, these technologies were used only for military and research purposes and then by technology enthusiasts and pioneer governments. It was only later when dynamic and innovative cities started involving these technologies. IT companies like IBM, Cisco, Siemens and General Electric started to employ their technology and investing in the improvement of cities (Barrera, 2020). Due to the widespread use of internet, computers, smartphones etc., there has been an enormous amount of data generated and even awareness; thus, citizens are also interested in deploying changes into a city to create a "smart city".

Over the years, the cities which incorporate technology into their management regulatory systems have been given many names. In 1987, the term *wired cities* (Dutton, Blumler, & Kraemer, 1987) was coined for a city with telecommunications between households and businesses. Later on, the concept of *cyber cities* (Graham & Marvin, 1999) was stated as a city with complex interdependence between transport and telecommunications, which link between urban and cyber cultures and economies and interdependently linked IT-based communities. Further, the term *digital city* (Ishida & Isbister, 2000) came into the picture and it "is defined as a city that integrates urban information (both achievable and real time) and creates public spaces for people living in the cities". In 2002, the term *intelligent cities* (Komninos, 2002) was referred to as "where the innovation processes meet the digital world and the applications of the information society". Next was *the sentient city* (Shepard, 2011), wherein it can hear and feel things but doesn't have any particular information; it can remember, correlate and anticipate using certain technology like sensors and algorithms. Now, coming to *information city* (Chourabi, et al., 2012), it collects information from localities and delivers it to the public via the internet.

The cities combined with the ICT technologies have been referred using various terms over the years, each with a different underlying concept, but share a focus on the effects of ICT on urban form, processes and modes of living. Therefore, all can be combined under an umbrella term of *smart city* which is the topic of concern in the chapter.

In 1955, a new concept of artificial intelligence (AI), coined by Dr. John McCarthy, emerged enabling to explore areas which were never thought of before. Its concepts consisting of machine learning, deep learning, computer vision, Internet of Things (IoT) etc. merged with the ICT technologies will bring revolutionary changes in the cities, aiming to become a smart city.

1.4 INTRODUCTION TO MACHINE LEARNING

The focus now shifts to a very critical concept and its need for building a smart city. However, first, to understand this, it is important to talk about "artificial intelligence".

Every species survives based on the type of intelligence they possess. *Homo sapiens* is termed the most "intelligent" species of all because their intelligence consists of many additional and more advanced qualities as compared to others.

These qualities are as follows:

1. *Estimation or prediction*: It is the quality of forecasting a result based on the factors affecting them.
2. *Differentiation or classification*: It is the quality of categorizing different items into definite classes, for example, cats and dogs.
3. *Segmentation*: It is the quality of grouping similar things together, for example, plants and animals.
4. *Recommendation*: It is the quality of recommending something to someone according to their interests.
5. *Decision-making*: It is the quality of choosing based on the information gathered.

Now, when a machine is made or taught to mimic the human tendency to fulfil the previous five qualities of intelligence, it is termed AI.

Shifting focus to the whole basis of the chapter, we can see that machine learning is a subset of AI. In simple words, it consists of algorithms or models which help in achieving AI. It is an intersection of the field of mathematics, computer science and optimization.

Two sub-parts of machine learning will be discussed further.

1.5 INTRODUCTION TO MACHINE LEARNING ALGORITHMS

As discussed earlier, machine learning helps achieve the various characteristics of AI. As the words suggest, it is all about the machines "learning". But what is meant by this? How is a machine supposed to learn? The answer to this is simple, through machine learning algorithms.

Machine learning algorithms are certain mathematical programs that are programmed in a way to provide basic guidelines to create a model. They are accustomed to adjust to the data provided by the user.

Now, before going further to understand the types of machine learning algorithms, it is important to first discuss some important terms which will be used throughout the chapter.

1. *Dataset*: It is a set of data, i.e., collection of data which can be in the form of tables, SQL queries, images and videos.
2. *Training dataset*: This dataset is used to train the machine learning algorithm to create a model.
3. *Validation dataset*: It is a part of the training dataset which is set aside to get an estimate of the performance of the model and tune the hyper-parameters accordingly.
4. *Testing dataset*: It is also a part of the training dataset, and its use is quite similar to the validation data except here the hyper-parameter tuning doesn't take place and gives the results of the final fine-tuned model.

1.5.1 Supervised Machine Learning Algorithms

Supervised learning is derived from the word supervisor who is a person "supervising" a particular task and providing guidance.

For example, when the students are first taught addition, they are provided with a method and some examples to understand the concept. When using this, they understand the task at hand; they are provided with questions similar to the examples to apply what they had learnt.

Similarly, in supervised learning, the machine is taught using some examples called a labelled *training dataset*, through a method or an algorithm. After this, the *validation dataset* is used to fine-tune the hyper-parameters to create an optimum model. When the machine is done learning, unknown data called the *test dataset* is introduced for the machine to provide an unbiased evaluation of a final model fit on the training dataset.

The supervised machine learning algorithms are categorized into two types:

1. *Regression*: This type of analysis is used to map the dependent variables to independent variables to predict the output in the form of continuous values such as a floating point or an integer.
2. *Classification*: In this type of analysis, the prediction output is in the form of discrete values such as classifying if a mail is a spam or ham.

Now, it is the time to discuss some algorithms which can be used to fulfil the tasks of regression, classification or both.

1.5.1.1 Random Forest

To first understand the random forest algorithm, it is imperative to know what are *Decision Trees* and define *Ensemble methods.*

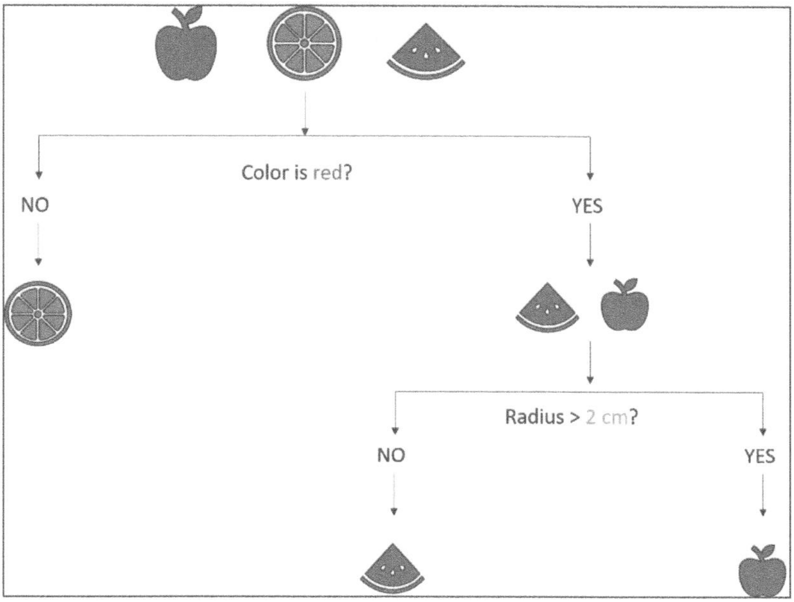

FIGURE 1.1 Decision tree example.

To get a clear idea, an example of a classification problem is provided (Figure 1.1). There is a basket of fruits having apples, oranges and watermelon slices, which need to be segregated into three different baskets with the same type of fruits. This can be done with the help of a decision tree as given in the following.

To get results from a decision tree, a feature split is done to move closer to the classification. So, to split our first node, the question "is the colour of the fruit red?" can be used. As one can observe, watermelon slices and apples are classified as red and orange as not red. Therefore, for oranges, it's clear how to classify them. The next node split can be done by answering the question "radius is greater than 2 cm". These watermelon slices cannot have a radius greater than 2 cm, therefore, are categorized accordingly.

Now, through the example mentioned previously (Figure 1.1), the contents of the fruit basket have been classified into oranges, apples and watermelon slices using two node splits. This was a really simple example to understand decision trees, but in reality, there can be a greater number of nodes used for prediction.

Next, the topic of conversation is *Ensemble Learning*. Sometimes, one type of algorithm cannot generate the desired level of accuracy and to solve this, predictions from models created by multiple machine learning algorithms are combined to improve the predictive performance.

Here, in a random forest, a particular type of ensemble learning is used, known as bagging. Using this method, many decision trees are used to make predictions on different samples of the same dataset and an average of these predictions is considered output for *Random Forest*. The main advantage of this widely used algorithm is reflected when there is a low correlation among the different decision tree models.

The main reason for this is, the lower the correlation, the higher the chances of individual errors of each decision tree not affecting the other tree.

Also, random forest can handle missing data well, robust to outliers, less impacted by noise and is extremely stable, which makes it a very useful algorithm to apply in a smart city.

1.5.1.2 Support Vector Machines (SVMs)

Now, to understand this algorithm, a classification problem statement is used. Along with which some relevant terms will also be defined.

There are two class points represented by an addition operator and a triangle as can be seen in Figure 1.2. In support vector machine (SVM) algorithm, one can classify these points using a hyperplane. Hyperplanes are decision boundaries used to classify data points. As can be observed in the diagram, the hyperplane classifies the two points into different classes. But the story doesn't end with just a simple hyperplane. To make the data points easily linearly separable, the algorithm also creates two margin lines parallel to the hyperplane having some distance from each other. These hyperplanes (on the margin lines) are not only parallel to the original hyperplane but also pass through one or more data points in their class. The main significance of creating these hyperplanes apart from the original hyperplane is to create a more generalized model giving comparatively accurate results. The whole distance between the two hyperplanes created is called the *Margin Distance* and the data points through which these planes pass are called *Support Vectors,* hence, the name, SVMs.

The example discussed earlier gives a basic intuition of how the algorithm works. In reality, one has to work with n-dimensional data which cannot even be visualized and can be non-linearly separable. Another example of non-linearly separable data to understand this better is shown in Figure 1.3.

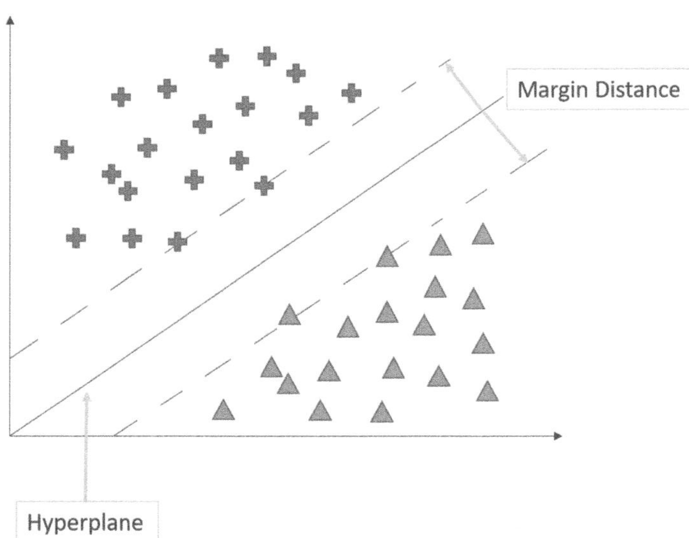

FIGURE 1.2 SVM example.

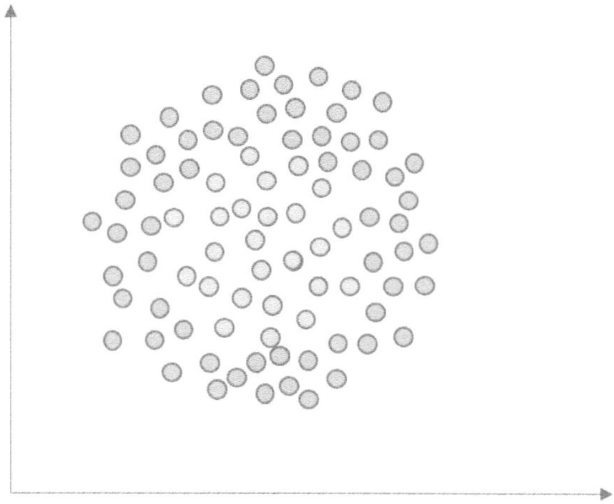

FIGURE 1.3 Non-linearly separable data.

If one treats the non-linear separable data given in Figure 1.3 as linearly separable and creates a hyperplane or line (as its 2D form), it will only give an accuracy of ≤50%. So, how to solve this problem?

To solve this, the SVM algorithm uses the SVM kernels. The SVM kernels convert low-dimensional data into a higher dimension which in this case is from 2D to 3D as shown in Figure 1.4.

As can be observed (Figure 1.4), a hyperplane can be easily created to classify the two types of data points with much higher accuracy than by just creating a line in its 2D form.

In smart cities, the data collected has multiple attributes which have to be processed to proceed to an accurate solution; therefore, SVM is a good choice in such cases as it performs profoundly well with higher-dimensional data.

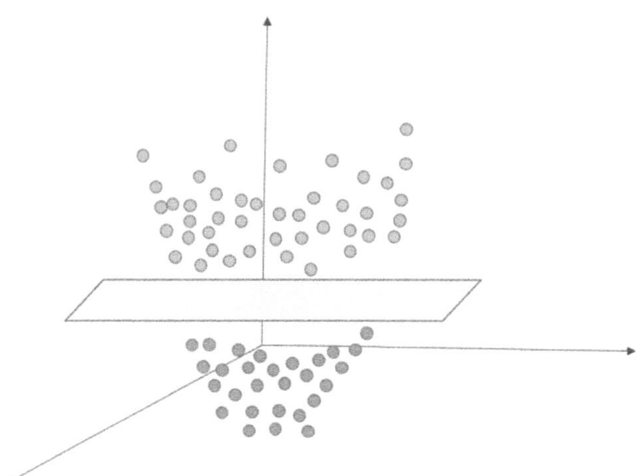

FIGURE 1.4 3D for non-linear separable data.

1.5.1.3 Naïve Bayes

To understand the Naïve Bayes algorithm, it is crucial to understand some statistical concepts called *Conditional Probability* and *Bayes' Theorem.*

Conditional probability is the one of an event A occurring such that event B has already occurred. The formula for the same can be observed as given in the following:

$$P(A|B) = \frac{P(A \cap B)}{P(B)}$$

Bayes' theorem gives the probability of an event based on the probabilities of the data provided in the hypothesis. It assumes that every input variable is dependent on all the variables. The formula for the same is as follows:

$$P(A|B) = \frac{P(B|A)P(A)}{P(B)}$$

The concept can be applied in an algorithm in the following manner. Let there be a set of inputs $X = x_1, x_2, x_3, \ldots, x_n$ and y be the output to be classified. The probability of x is known and the probability of y is to be calculated. This can be represented as follows:

$$P(y|x_1 \cap x_2 \cap x_3 \cdots \cap x_n) = \frac{P(x_1|y) * P(x_2|y) * P(x_3|y) \cdots * P(x_n|y) * P(y)}{P(x_1) * P(x_2) * P(x_3) * \cdots * P(x_n)}$$

Just as observed in Bayes' theorem, the event which has occurred has been replaced with the input features x.

Now, assume that the event has to be classified as True or False and by the algorithm formula applied earlier, the resulting probability is 0.8 for True and 0.2 for False. To get the results, one more formula has to be applied, which will give the final classification result:

$$y = argmax\left[P(y) \prod_{i=1}^{n} \Pi(P(x_i) \mid y) \right]$$

where the argmax operation returns the result with the maximum probability, which in this case is 0.8 to classify the set of features as True. Hence, this is the intuition behind the working of the Naïve Bayes algorithm.

The Naïve Bayes algorithm is quite useful in a smart city setup as it is very simple to implement. Also, it requires a less amount of training data to train, thus making it a computationally fast algorithm.

1.5.1.4 K-Nearest Neighbours

This algorithm is explained using two use cases, classification and regression.

First focus of attention is classification. As can be observed from Figure 1.5, the data points belong to two classes, blue and green. A new data point indicated by the red star has been introduced, and it has to be classified in either blue or green class. But the question here is, how to solve using the K-nearest neighbours (KNN) algorithm?

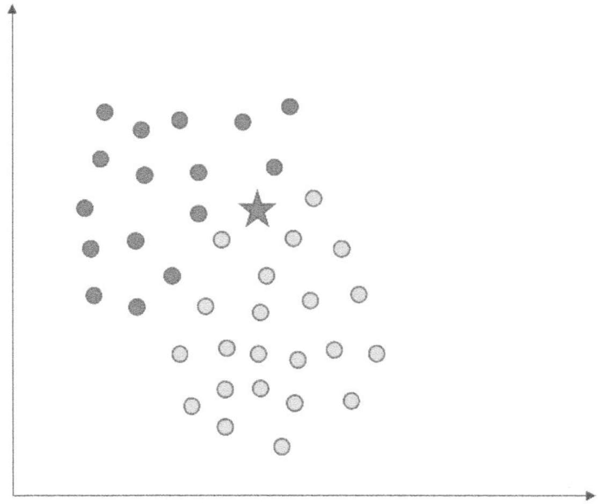

FIGURE 1.5 KNN classification.

First, the value of K must be chosen (assumed to be 5). The next step is to find K's nearest neighbours, i.e., five to the point to be classified. These five points are determined using the Euclidean distance between two points whose formula is as follows:

$$d = \sqrt{(x_2 - x_1)^2 - (y_2 - y_1)^2}$$

After the algorithm calculates the distance, it concludes that out of the five nearest points, two belong to the blue class and three belong to the green class, which leads to the conclusion that the new point belongs to the green class.

The second order of business is learning how to solve a regression problem.

In regression, the floating or integer value of a point, given some attributes, is to be calculated. Let's say a person is provided with a task to find the value of the point represented as a star in Figure 1.6. Just like before, they are required to take the value of the nearest neighbours, i.e., K = 5. Now, the algorithm will calculate the mean of the five nearest values and assign that value to the unknown star point to be predicted.

KNN is also called a *Lazy Learner* as it is based on instance-based learning, i.e., it does not learn or train during the training period. It essentially stores the training data and uses it only at the time of making real-time predictions. Due to this, it's an efficient algorithm and is much faster than other algorithms that require training. Therefore, it is a good choice to implement in a smart city background.

Summarizing the type of tasks each algorithm can perform:

Algorithm	**Classification**	**Regression**
Random forest	✔	✔
Support vector machines (SVMs)	✔	✔
Naïve Bayes	✔	✘
K-nearest neighbours (KNN)	✔	✔

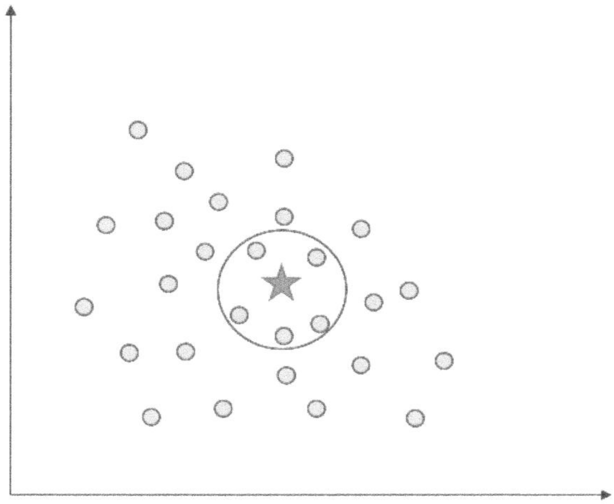

FIGURE 1.6 KNN regression.

1.5.2 Unsupervised Machine Learning Algorithms

In *Unsupervised Learning*, no guiding force is present to train the machine and the data provided is unstructured and unlabelled on which the algorithm is allowed to act upon. The machine must identify patterns in the data by itself. The main goal is to find the similarities and differences between data points.

For example, a dataset contains an unlabelled image dataset of three fruits – apple, banana and strawberry. The machine cannot identify the type of fruit by looking at the picture. Without any labels guiding it, using an algorithm, the machine has to identify different features, hence categorizing them into clusters.

The discussion now moves on to an important unsupervised learning algorithm.

1.5.2.1 K-Means Clustering

K-means clustering is an unsupervised learning algorithm and is thus introduced to unlabelled data. It is used to detect relationships between data by clustering data points with similar characteristics together (Figure 1.7). When the clusters are formed, it is upon the data scientist to identify the behaviour on which the division has been created. It can uncover interesting facts about different things which the data might be hiding, for example, the age of people, income bracket and geographic location.

Now, let's discuss how the algorithm works. First, one needs to randomly assign the value of K, i.e., the number of clusters. There is no definite answer for the number of clusters that should be taken as there might be a change in what each represents. After deciding the number of clusters, assign each of them a centroid. A centroid may be an imaginary or real location of the centre of the cluster. Each data point is assigned to a cluster, the centroid of which is closest to it. The main objective of the K-means algorithm is to minimize the sum of the distances between the data points and the cluster centroids. Therefore, the initial random position of the centroid is adjusted through several iterations to create optimum clusters of data points.

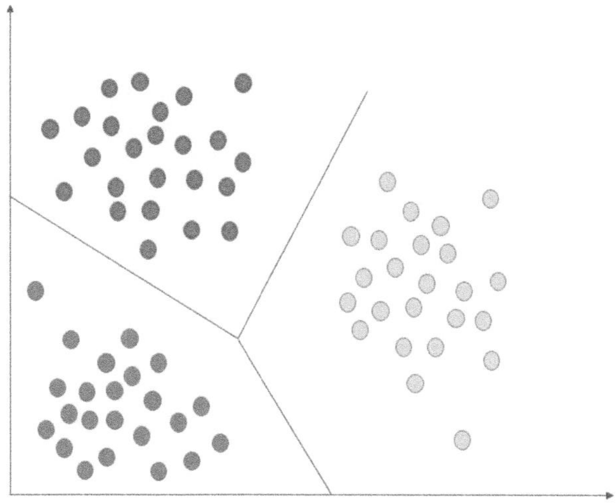

FIGURE 1.7 K-means clustering.

However, there are some drawbacks to this unsupervised learning algorithm.

a. *Choosing the value of K is a tricky task*: One way to solve this problem is to keep increasing the number of clusters till one stops getting large reductions in the distance between the data points of the centroid of the clusters.
b. *Avoiding the local minima*: As discussed earlier, the algorithm has to minimize the sum of the distances between the data points and the clusters. When doing so, it may converge to a local minimum; thus, several iterations have to be done to avoid that.
c. *Assigning the data labels*: When the optimum clusters are formed, the data scientist has to identify the characteristics based on which they were created. This is difficult, as there can be several meanings associated.

K-means clustering is a quite flexible algorithm as it can easily adjust to changes like cluster segment. It is also suitable for large datasets and as smart city systems have huge amounts of data to configure, K-means can prove to be useful.

1.5.3 Neural Networks

1.5.3.1 Artificial Neural Network (ANN)

When discussing AI, the most significant point about it which comes to mind is the computer's ability to mimic human behaviour. In humans, the brain and the central nervous system take care of doing innumerable actions like making decisions, walking and learning. Therefore, to make computers act intelligently, one has to create a structure similar to that of the brain and the nerves, the main component of which is a cell called *Neuron*.

Roughly, a biological neuron consists of dendrites, nucleus and axon (Figure 1.8). Dendrites are the input terminals that send the information for processing into the

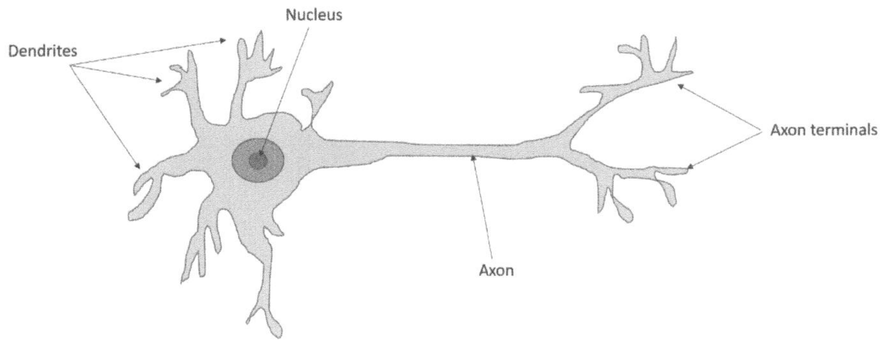

FIGURE 1.8 Biological neuron.

nucleus, which makes a decision and sends the output signal through the axon. These signals from the axon can also act as input for the subsequent neurons as millions of them are connected. Mathematically, this neuron can be perceived as given in Figure 1.9. This is known as the *Perceptron Model.*

The inputs which are fed into the neuron are assumed to be *x_1* and *x_2* and the weights to be *w_1* and *w_2* (Figure 1.9). An equation is formed to get the output *y* and the weights are the coefficients of the equation:

$$w_1 * x_1 + w_2 * x_2 = y$$

All the inputs cannot be of equal importance or “weight” to the outcome; thus, some particular value called *weight* is multiplied with the input to amplify or condense its effect on the output. However, sometimes the value of input can be 0 and whatever weight it is multiplied with also amounts to 0. So, a term called *bias* is introduced to avoid this:

$$(w_1 * x_1 + b_1) + (w_2 * x_2 + b_2) = y$$

Then, the previous aggregation of inputs is passed into an activation function. Activation function is the one that provides us with an output of the neural network

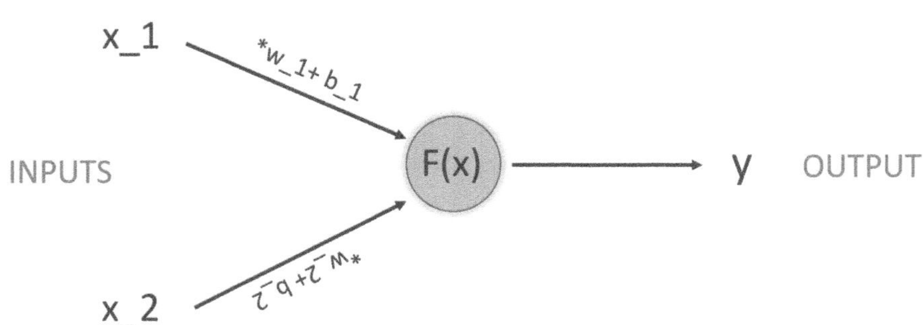

FIGURE 1.9 Perceptron model.

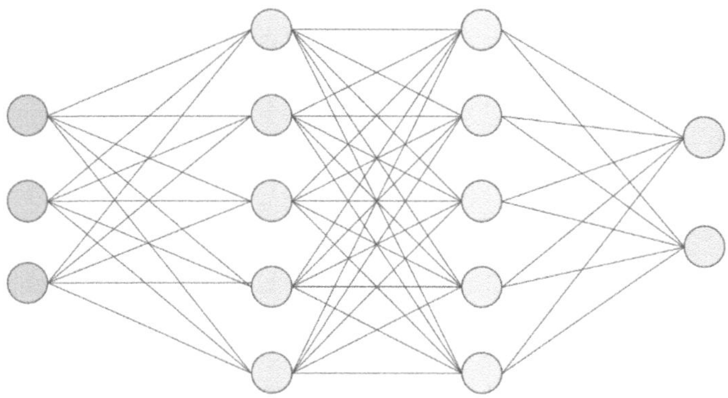

FIGURE 1.10 Multi-layered perceptron model.

like True or False, a range of data between 0 and 1 or –1 and 1 etc. depending upon the type of function:

$$\varnothing_a\left(\left(w_1 * x_1 + b_1\right) + \left(w_2 * x_2 + b_2\right)\right) = y$$

Now, just as in the brain, neurons are connected, perceptrons can also be connected to create multiple layers of perceptrons which is known as a *Multi-Perceptron Model* or a *Neural Network* (Figure 1.10). There are several types of neural networks like recurrent neural network (RNN) and convolutional neural network (CNN) which will be discussed in the upcoming sections of the chapter. In this section, a feed-forward Network is talked about, where every neuron in each layer is connected to every neuron in the next layer, as can be observed in Figure 1.10.

This artificial neural network or ANN can be used to solve both regression and classification problems. They consist of three types of layers:

a. *Input layer*: This is the layer that feeds the data values into the network.
b. *Hidden layers*: These layers are between the input and output layers where weights and biases are applied and entered through an activation function.
c. *Output layer*: This is the final layer that gives us the final outcome and can consist of one or more neurons.

The input and output layers are quite straightforward and simple but it is difficult to understand what is going on in the hidden layers in terms of the weights, biases. Due to such a complicated network, networks with more than two hidden layers are known as *Deep Neural Networks*.

ANNs have a high accuracy with fault tolerance, i.e., even if one or more neuron of the network is corrupted, it doesn't let it hinder the process of output generation. This makes the ANN a good choice to be applied in the making of a smart city.

1.5.3.2 Convolutional Neural Network (CNN)

CNN is a neural network used to process image data for image classification, for example, classifying the images into cats and dogs and identifying the digits. This

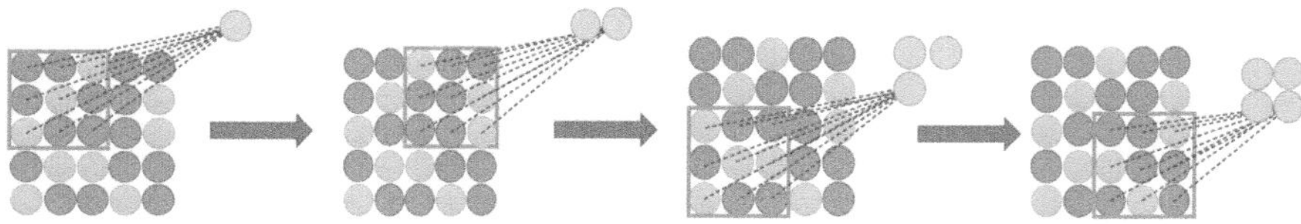

FIGURE 1.11 Convolutional layer.

image classification task can also be carried out by ANNs, but the results will be quite poor. This is because of the following reasons:

a. In ANN, every neuron in a layer is connected to every neuron in the next layer, which would create millions of parameters to process and it would make the computation difficult.
b. One will have to flatten the 2D image pixels into 1-D array as ANNs can only compute 1-D data and this would lead to data loss giving inaccurate results.

In CNN, there is local connectivity, as all the neurons in a layer are not connected to every neuron in the next layer, creating fewer parameters and only focusing on the important features. This neural network has one other major advantage, image filters. Image filters have a height and width smaller than the original image and are made to stride across it to highlight certain features which are used to provide accurate results. For example, it can be used to detect edges, vertical or horizontal lines etc.

There can be hundreds of filters in a layer that is known as a *convolutional layer* which is created as follows for a grey-scale image.

As observed from Figure 1.11, stride an image filter over the original image which here has a small dimension of 5 × 5. Note that the stride size can be increased and doesn't necessarily have to be one. By striding, localized connections are created between the input and the next layer which essentially creates a filter. Now, the process can be done for any number of filters as shown in Figure 1.12.

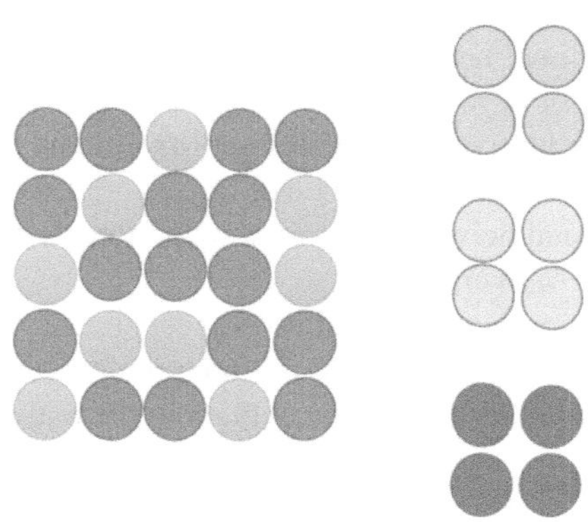

FIGURE 1.12 Multiple-image filters.

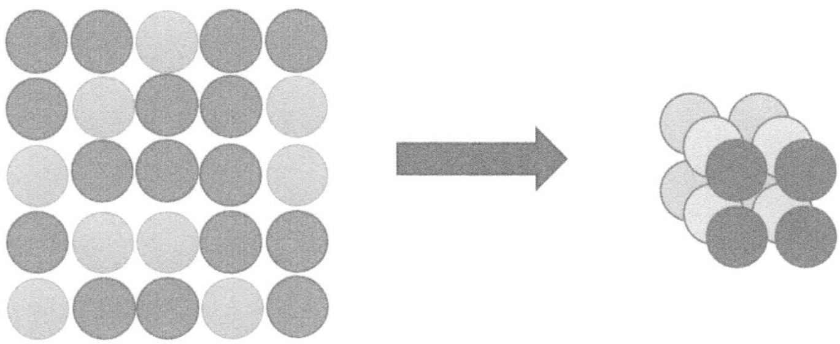

FIGURE 1.13 Convolutional layer.

The number of filters depends upon the complexity of the image to be classified. The filters are then stacked to create a convolutional layer as can be observed in Figure 1.13. This layer has localized connections.

The process discussed earlier is for the grey-scale images, but real-life experience may provide with coloured images. Coloured images are created using the primary colours: red, green and blue; thus, the image has three colour channels, which makes it 3D. Due to this, a 3D filter is created, one for each channel.

Now, although the convolutional layers have localized connections, there can still be a lot of parameters to process. Due to this, the process of pooling is conducted. *Pooling* is another building block of CNN that is used to reduce the spatial size of the representation (Figure 1.14).

This process of image classification has various applications which will be discussed in the later sections.

CNN has the computational ability to handle enormous amounts of data for prediction. One doesn't have to perform a lot of pre-processing to feed the data into the network for accurate results, thus being an optimal choice for image processing in a smart city.

1.5.3.3 Recurrent Neural Network

Till now the discussions have been around regression and classification problems and solving them using different algorithms. But these are not the only applications of machine learning and deep learning as they can also deal with a sequence of data.

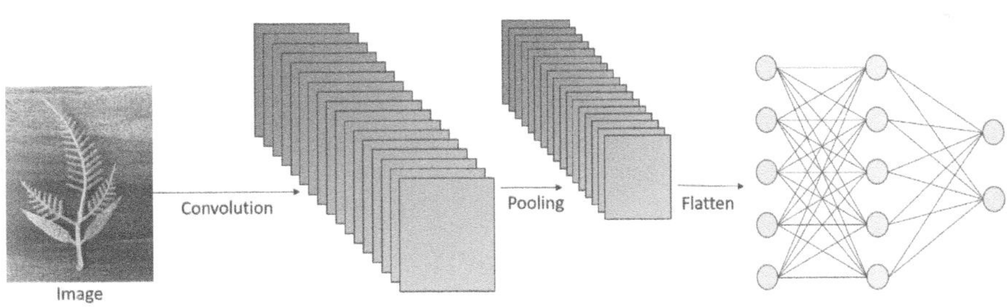

FIGURE 1.14 Convolution neural network.

FIGURE 1.15 Time series example.

The sequence of data refers to the dataset of points wherein they are dependent on the previous points. For example, in time series data (Figure 1.15), to predict the sales of the upcoming months, one has to analyse the pattern of the sales in months of the previous years.

RNN is a type of neural network which deals quite elegantly with this type of data and is powerful enough to give good results. The working of this network is not very complicated. In feed-forward networks, an aggregation of inputs is performed and then passed through an activation function to get an output (Figure 1.16).

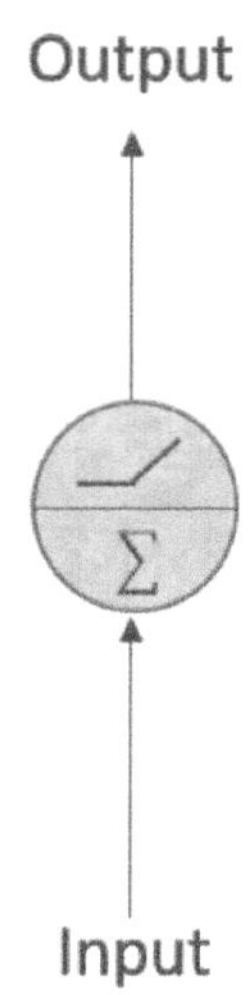

FIGURE 1.16 Feed-forward neuron.

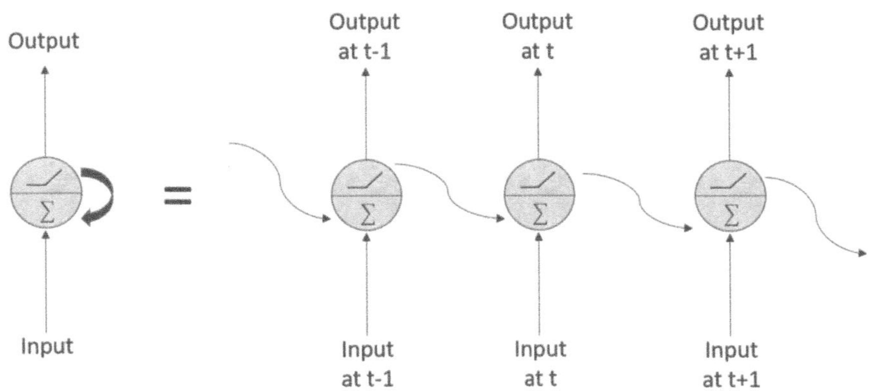

FIGURE 1.17 RNN process.

Similarly, in RNN (Figure 1.17), an aggregation of inputs is performed and passed through an activation function to get an output. But due to dealing with a sequence of data here, the previous output has to be inputted along with the input to the next neuron to get the subsequent output and so on.

RNN can also be used in natural language processing or NLP which is a branch of AI dealing with human linguistics. Its applications are chatbot, sentiment analysis, Email filters, predictive texts etc. and use RNN for many of these.

RNNs as explained not only take the general input but also the output of the last cell and thus remember every information through time. It is also computationally efficient in a smart city for predicting time series data, NLP, and also can be used to improve CNN.

1.5.4 Computer Vision

In the era of AI, the significance of real-time image processing and computer vision has increased exponentially. *Computer Vision* is used to derive useful details from digital images, videos and other visual inputs. It has a variety of examples and also important applications in smart cities.

To learn what is object detection, it is important to first understand *Object Localization*. In this image recognition process, one has to locate an instance of a particular object in an image as shown in Figure 1.18.

Although, in object detection, one has to locate various objects in a single frame, similar to object localization in Figure 1.18, a bounding box is created around the object to detect it (Figure 1.19).

Computer vision has a huge set of new applications such as the training of self-driving cars and smart farming. It is a new area that has developed immensely in just a few years and is more widely available to explore. Earlier, it was a tedious task to set up an object detection model using the classical algorithms, but now there are libraries like OpenCV to make this task easy and simple. However, the rapid adoption of deep learning over the years has brought into light highly accurate object detection algorithms R-CNN, Fast-RCNN, Faster-RCNN, RetinaNet, SSD and YOLO.

FIGURE 1.18 Object localization.

Computer vision techniques perform real-time analysis and have a high reliability due to the absence of human factor of lethargy. Also, it has a high accuracy due to which there is a reduction of cost and time, giving way to spend the resources in another direction in a smart city environment.

Till now, various machine learning and deep learning techniques have been explored. In the next sections, the applications of these algorithms in a smart city set-up are talked about.

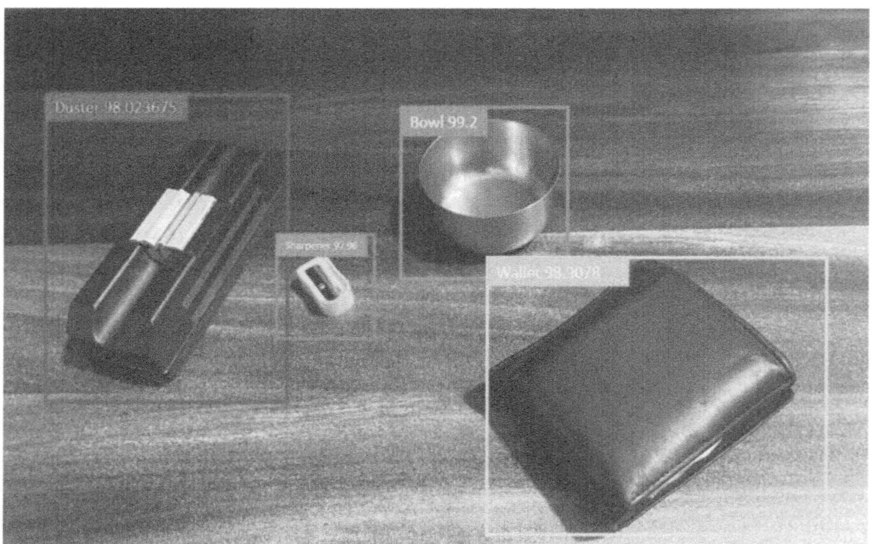

FIGURE 1.19 Object detection.

1.6 APPLICATIONS OF MACHINE LEARNING IN SMART CITIES

The machine learning and deep learning algorithms discussed in the previous sections were to bring a clarity of concept. The focus of this section is on the foundation of applications at the core of a smart city.

1.6.1 Traffic Management

In the last 15 years, both the demand and supply of automobiles have increased exponentially, thus gradually increasing road congestion. Traffic not only consists of personal vehicles but also cycles, bikes, trucks and public transportation like buses, autos and rickshaws. This tends to create a lot of raucous on roads, which is one of the major concerns to tackle, while planning a smart city. During some instances, citizens even tend to violate the traffic laws but go undetected leading them to violate the laws again.

Now, to solve these issues, smart cities integrate machine learning algorithms (Figure 1.20) into the systems. Due to the increase in congestion on roads as discussed earlier, people get stuck in traffic for long durations and are looking for time ranges to avoid during the day, as well as the optimal route to take. One algorithm which can be used for the *Traffic Congestion Prediction* (Yunxiang Liu, 2017) is random forest because of high robustness, high performance and high practicability. It will be a classification problem that takes into account many variables like weather conditions, time period, road quality and holidays. This is also similar to *Traffic Condition Recognition* for which an unsupervised learning algorithm such as K-means clustering can be implemented.

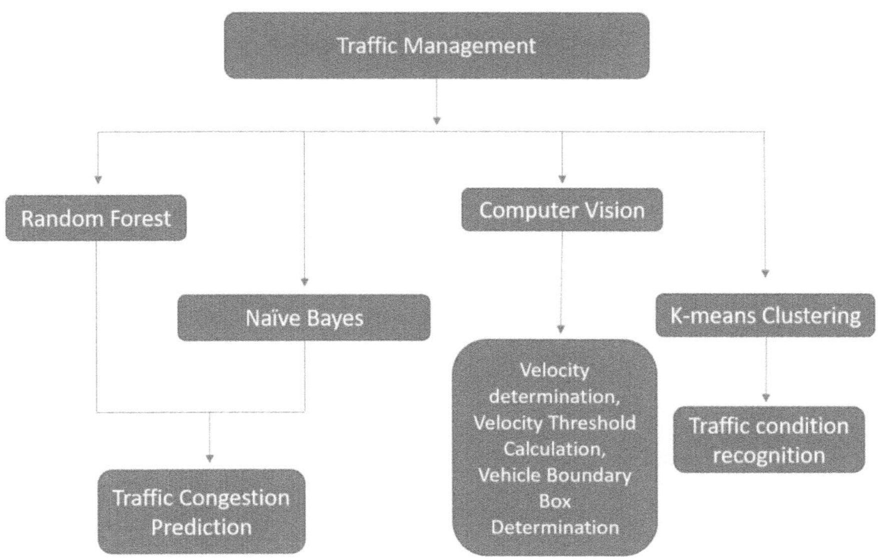

FIGURE 1.20 Traffic management.

Next is the problem of traffic rules violation which is tricky yet interesting to solve using computer vision. It can be used for velocity determination and its threshold calculation along with detection of all the vehicles on the road using vehicle boundary box determination.

1.6.2 Transportation

This application is an extension of traffic management systems laid out in a smart city. An integral part of such cities is not only micromanaging and optimizing the traffic and transportation systems but also the needs of its citizens and helping them make informed decisions. Public transportation system is an integral part of any city because it helps reduce pollution, decrease road traffic, improve community mobility and commuters' productivity. Thus, having a fare management system and customer satisfaction analysis is of utmost significance to create a successful smart city.

Prediction of future mobility behaviour through crowdsourced data (Semanjskia, Lopez, & Gautama, 2016) can facilitate the management of a smart city's mobility system while simultaneously ensuring the timely provision of relevant pre-travel information to its citizens. The forecasting model which can be implemented here is SVM (Figure 1.21).

Next, buses are one of the most used modes of public transportation in almost every part of the world; therefore, it becomes important to create an intelligent bus management system. A QR code system can be implemented where the person can enter their destination details to get per person fare and the availability of buses during that time. These can also be used as attributes to input into the Naïve Bayes algorithm to predict the probability of buses required in a particular route, a single

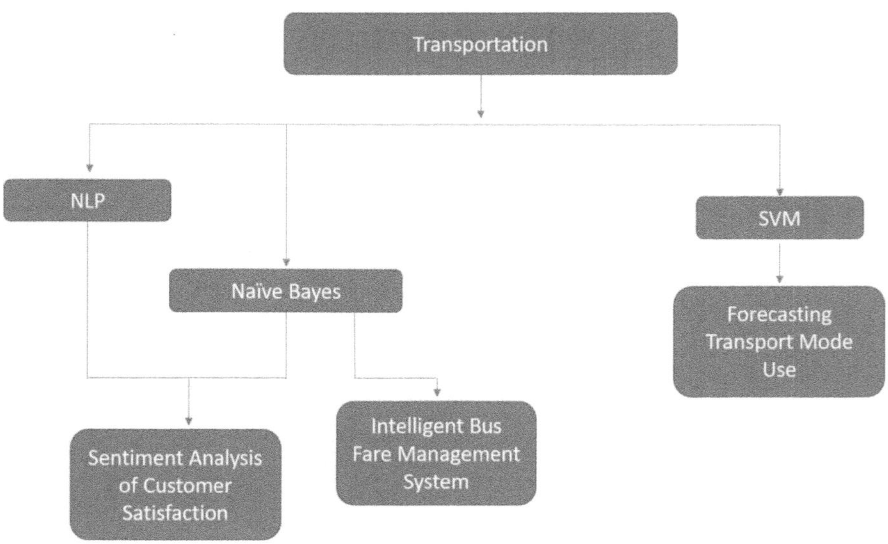

FIGURE 1.21 Transportation.

person travelling number of times in a day etc. and help the public transport companies in increasing the usage while reducing the operational cost (Maniraj, Sharma, Ratnam, & Kumar, 2020).

In any industry, customer satisfaction is given utmost priority and the same can be implied about the different transportation companies, both public and private. Sentiment analysis can be performed on the reviews posted online on the website or the social media handles and is beneficial to extract people's opinions and emotions to improve the services. This can be done using Naïve Bayes and NLP.

1.6.3 Healthcare

Any country's economic standing is predominantly dependent upon its healthcare system. Health is the most important aspect of a human's life and needs to be well maintained as only a thriving person can live a quality life and contribute to the economy. Back in time, the death rate was high due to less knowledge about diseases, how to detect them accurately and the appropriate treatment. Due to the gradual increase in the doctor's knowledge as well as the advent of technology, diagnosis has become easier and faster leading to timely treatment. Therefore, a smart city needs to have an efficient, top of the class yet equitable healthcare system.

Image classification techniques (Figure 1.22) such as CNN have proved to be accurate in disease classification. Images of X-rays for predicting pneumonia or classifying cells into benign or malignant for predicting cancer are some of the applications of the image classification techniques. The only issue with this type of classification is that of data labelling as it requires a lot of professional expertise which makes the datasets hard to collect.

IoT has become prevalent in the AI industry and combining it with machine learning algorithms takes the technological transformation to new heights. For healthcare, physical and mental health monitoring systems based on IoT can be set up which can be accessed even by the less privileged. The patient's data can then be stored on the cloud, and predictions can be made using various machine learning algorithms

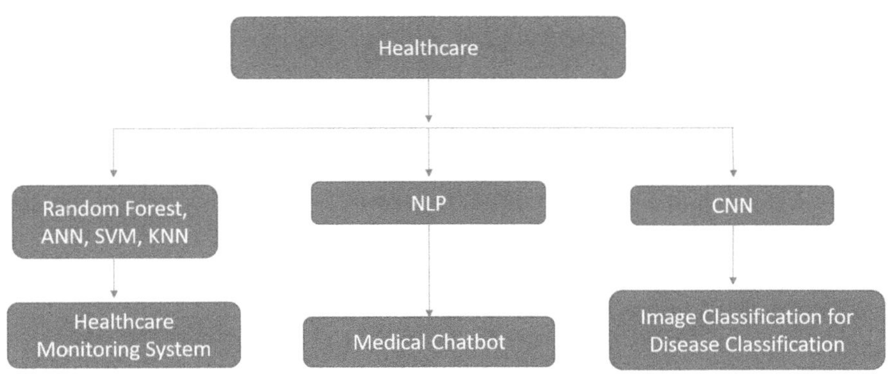

FIGURE 1.22 Healthcare.

like random forest, SVM, KNN, ANN to diagnose if the person is suffering (Kaur, Kumar, & Kumar, 2019).

Numerous patients have doubts, want to virtually book an appointment, order medicines directly from the hospital's pharmacy and book a blood test session and other medical services. This can be done by opting for a chatbot service which is created using machine learning techniques and NLP. Chatbot is a software application that enables one-to-one interaction to perform tasks through text or text-to-speech in place of directly contacting a live agent. The benefits of including one are to increase customer engagement, reduce customer service costs, monitor consumer data to gain insights, 24 × 7 availability etc.

1.6.4 Public Security

It is the responsibility of the government to ensure that the citizens are protected from the threats that pose in any society. This not only refers to the physical harm but also cyber security of the public like safety from data breaches, unauthorized exploitation of systems, phishing and password attacks. In public places, security personnel are available like the guards with arms for protection as well as police helpline numbers in any kind of distress. But sometimes, it becomes impossible to spot the culprit in a heavily crowded place like railway stations, metros, airports and malls, and the case goes unnoticed. Due to not getting caught, the malpractice of robberies and security breaches increase.

For detecting misconduct such as robbery, burglary, stalking and kidnapping, the installation of surveillance cameras in every nook and corner of a public place or a shady place is the first and one of the most important steps. Also, many times suspicious objects which pose a threat to the public at large might go undetected leading to an extremely serious threat. To notice such kinds of situations and alert the security personnel immediately will require a computer vision object detection model. Multiple bounding boxes will be present detecting live footage for any wrongdoing, even threats to women, child and senior citizens. Also, this model will be trained to accurately identify an unattended or unknown object and finally give a warning.

With the advent of technology, cyber threats and attacks have become much more prevalent and dangerous. This threat is not only for the general public and VIPs but also the whole smart city system itself. If there is a data breach in the city's system, it will make the smart city's functions unoperational and will lead to a huge privacy breach. Therefore, creating a strong anti-virus and other malware protection software is important to avoid financial and moral blackmail. This software will be present to avoid breaches, whereas the inclusion of machine learning algorithms (Figure 1.23) such as SVM and random forest will be for the detection of the data breach entering the system.

1.6.5 Energy Conservation

There has been a tremendous increase in fuel level consumption over the past two decades since the gradual lifestyle changes. The traffic on roads due to the increase in personal vehicles not only is a point of concern from the management perspective

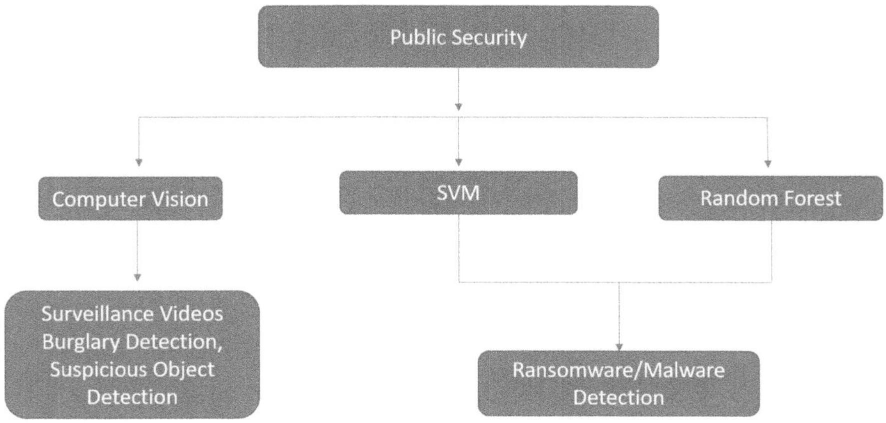

FIGURE 1.23 Public security.

but from the conservation as well. Along with vehicles, there is a huge demand for coal to generate electricity in households and big corporations like company buildings and factories. As smart cities' main aim is to sustainably increase the level of living of their citizens, energy conservation becomes of utmost importance.

The first step from the machine learning point of view (Figure 1.24) is to build an energy management system by creating an energy consumption forecasting for buildings and households. This is of significance in decision-making for effective energy saving and development in particular places. SVMs and ANNs can be

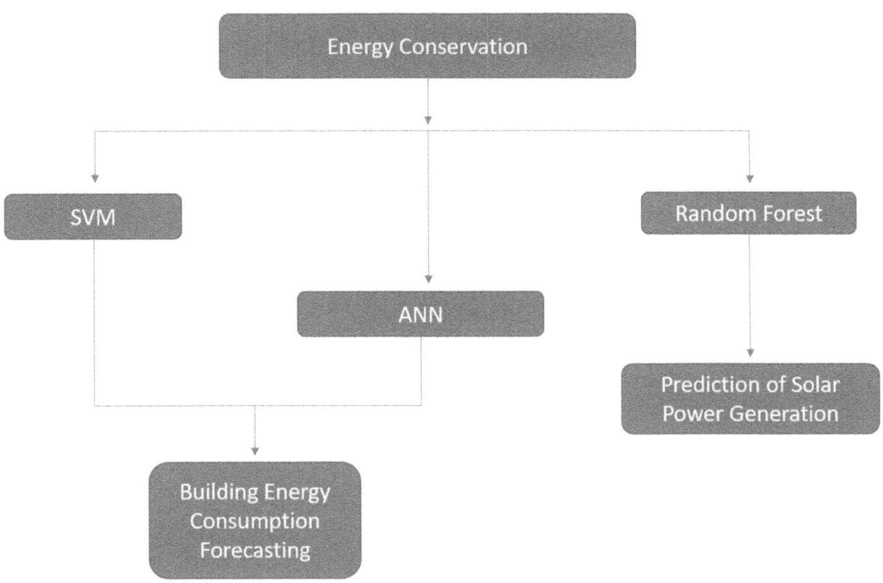

FIGURE 1.24 Energy conservation.

applied for this purpose as they have an ever-growing aim of finding the most accurate approach.

As an alternative to the non-renewable resources which is the main source of energy, setting up renewable sources of energy such as solar panels is an expensive yet effective method. Other sources such as wind energy and tidal energy require huge land mass and an extreme amount of funding which is impractical and not possible to set up in a city. Therefore, solar energy becomes the primary source of alternative energy for a smart city. Thus, the random forest algorithm can be used for the prediction of the amount of solar energy which can be used for power generation and distribution on a single day.

1.6.6 Pollution Control

This application is also linked with the increase in vehicles and factories causing unbearable amounts of air, water and noise pollution. Controlling pollution is the need of the hour to create a sustainable and healthy living. There are also cases of environmental pollution through forest fires and active volcanoes, but here the focus will be man-made pollution and how it can be curbed in a smart city.

In weather system applications and other softwares, the calculation of air quality index has become quite common. This air quality index is used to alert the public when air pollution comes to hazardous levels. Using machine learning, two applications can be created to curb air pollution using this index. First, the index is prone to various errors which can be improved using algorithms like random forest. After getting accurate results, a database for the prediction of air quality index can be created and can be solved using SVMs, random forest, and ANN (Figure 1.25). The importance of this prediction is that a well-planned framework can be created beforehand to deal with the situation.

Similarly, water quality analysis can be performed using unsupervised learning methods such as K-means clustering and traffic noise prediction using RNNs. The significance of these is the same as for the air pollution, i.e., giving the concerned legislation bodies some data to make plans.

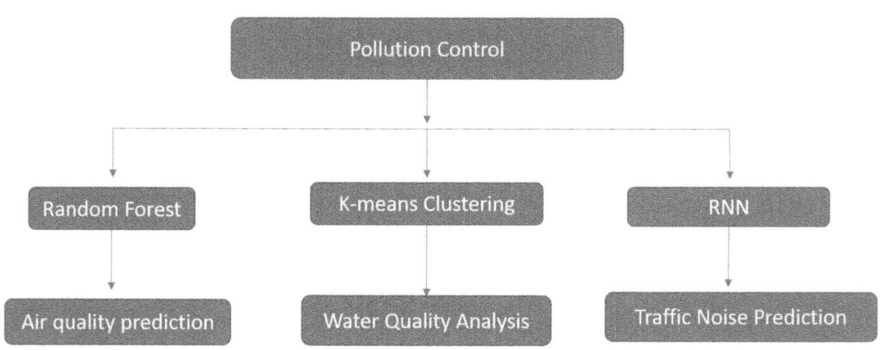

FIGURE 1.25 Pollution control.

In the next section, let's discuss the challenges of the implementation of the applications which were discussed earlier.

1.7 CHALLENGES OF IMPLEMENTATION IN SMART CITIES

1. Privacy Concerns
   Smart cities require real-time monitoring of the citizens in maintaining law and order. However, it has the potential to infringe upon the basic rights to privacy of the citizens. Real-time monitoring may lead to unsolicited tracking of the citizens. Data must be collected legally, responsibly and for the intended uses and purposes. This requires stringent laws and regulations. Policies must be laid down to ensure that
   - The basic rights of the citizens are maintained.
   - The monitoring data cannot be accessed for illegal activities or fall prey to the wrong hands.
2. Cybersecurity Risks
   In continuation of the privacy concerns, the data and the technological systems used must have sufficient technological security around them. Firewalls and checks must be built to ensure that unethical hacking or unlawful access to data is prohibited. Constant monitoring of threats and due diligence are necessary to ensure that the data is kept safe and cannot be used for harmful activities.
3. Data Capacity
   As there is real-time monitoring of data from security cameras, drainage systems, traffic lights etc., one can imagine that the many gigabytes of data be processed and transmitted every second. For the smart city to not collapse, there must be enough data and network centres to collect, store and process data.
4. Efficient Real-Time Data Processing
   Data processing must be instantaneous and real time for the efficient management of operations. High-speed internet connectivity is thus a necessity rather than a perk. Both the software and the hardware must be in tandem to ensure smooth conduct of affairs.
5. Natural Damage/Theft of AI Equipment
   Technological equipment, such as security cameras, traffic lights and sensors, must be kept safe and well-guarded. They must work smoothly under all weather conditions and not take huge weather damage. Moreover, they must be kept well hidden from the public view to minimize public encroachment.

1.8 FUTURE DIRECTIONS

As established in the text, AI technology is the way to go to transform a city into a smart city. The primary issue which needs to be tackled here is how to prepare the technology to thrive under certain conditions along with the monetary stand of the country. Huge amounts of investment are required to move into this sustainable yet technologically efficient environment. Another complication which needs to be handled to be successful in establishing a smart city is data. Data is the foundation on which smart cities are aimed to be created and are to be handled by experts. The concerned personnel should be equipped with managing large volumes of data and deliver real-time insights. Also, this collected data should be accurate and consistent with valuable inputs from the households and businesses to understand their needs.

Cities are complex places and the new technology which is being introduced into the cities is to be blended in accordance with the existing infrastructure, i.e., the perfect blend between the city's past and future. Smart cities emerged for a reason to make the lives of the citizens and the environment they live in better. Therefore, these only have a future if they digitally transform lives by solving real problems creatively.

1.9 CONCLUDING REMARKS

To conclude, AI backing up the smart city systems can completely change the lifestyle of citizens for the better. Throughout this chapter, the various reasons were put forward for which smart cities are being developed and are the future if governments are able to make the implementation a success. Also, there was a discussion about machine learning and its algorithms, both supervised, unsupervised and the deep neural networks so as to have a clarity of concepts before understanding the applications. Smart cities will change the face of sustainable living and create better societies for its citizens to enjoy.

REFERENCES

Barrera, J. O. (2020). *A Historical View of Smart Cities: Definitions, Features and Tipping Points*. Montreal: UQAM.

Chourabi, H., Nam, T., Walker, S., Gil-Garcia, R., J., Mellouli, S., …, & Scholl, H. J. (2012). *Understanding smart cities: An integrative framework*. In 45th Hawaii International Conference on System Sciences.

Dutton, W. H., Blumler, J. G., & Kraemer, K. L. (1987). *Wired Cities: Shaping Future Communication*. New York: Macmillan.

Graham, S., & Marvin, S. (1999). Planning cybercities: Integrating telecommunications into urban planning. Town Planning Review, 70(1), 92.

Ishida, T., & Isbister, K. (2000). *Digital Cities: Technologies, Experiences, and Future Perspectives*. Berlin: Springer.

Kaur, P., Kumar, R., & Kumar, M. (2019). *A Healthcare Monitoring System Using Random Forest and Internet of Things (IoT)*. Berlin: SpringerLink.

Komninos, N. (2002). *Intelligent Cities: Innovation, Knowledge Systems and Digital Spaces*. London: Routledge.

Maniraj, S., Sharma, A., Ratnam, I., & Kumar, A. (2020). Intelligent bus fare management system using Naive Bayes algorithm. IJRTE, 8(6), 1.

Semanjski, I., Lopez, A. J., & Gautama, S. (2016). Forecasting transport mode use with support vector machines based approach. *Transactions on Maritime Science*, 5(2), 111–120.

Shepard, M. (2011). *Sentient City: Ubiquitous Computing, Architecture, and the Future of Urban Space*. Cambridge: MIT Press.

Yunxiang Liu, H. W. (2017). *Prediction of Road Traffic Congestion Based on Random Forest*. New Jersey: IEEE.

2 Smart Cities in the Age of AI and the Internet of Things

Commitments and Obstacles

Mukta Goyal[1], *Ekata Gupta*[2], *Abhishek Srivastava*[3], *and Prabha Kiran*[4]
[1]Guru Nanak Dev Institute of Technology, Delhi, India
[2]Guru Nanak Institute of Management, Delhi, India
[3]Gopal Narayan Singh University, Sasaram, Bihar, India
[4]Westminster International University, Tashkent, Uzbekistan

2.1 INTRODUCTION

In light of the quick increase in population density inside metropolitan conditions, bases and administrations have been expected to address the issues of residents. Subsequently, there has been a striking expansion in the quantity of computerized gadgets, such as sensors, actuators, cell phones, and shrewd machines, which drive the tremendous business targets of the Internet of Things (IoT), in light of the fact that all gadgets can be interconnected with each other through the Internet. It was beforehand troublesome, whereas, on the off chance, it is not difficult to combine these advanced gadgets. In this chapter, we'll assess the capacity of shrewd water frameworks as well as how they can assist us by preserving this significant asset. Savvy urban areas are an ideal option compared to public arrangements and advances of the previous century, which are not yet equipped for managing momentum concerns, for example, population increase, water security, and environmental changes.

In light of the quick increase in population density inside metropolitan conditions, bases and administrations have been expected to address the issues of residents. Subsequently, there has been a striking expansion in the quantity of computerized gadgets, such as sensors, actuators, cell phones, and shrewd machines, which drive the tremendous business targets of the Internet of Things (IoT), in light of the fact that all gadgets can be interconnected with each other through the Internet. It was beforehand troublesome, whereas, on the off chance, it is not difficult to combine these advanced gadgets. Also, gathering data for everyday tasks of the board and long haul advancement requires to design the city. Some open transportation information, for example, use of constant area, parking spot inhabitance, gridlocks, as well as various information, such as weather patterns, air and commotion contamination status, water defilement, and energy utilization, ought to be gathered consistently.

DOI: 10.1201/9781003225317-2

With that in mind, different innovations have been used to address the extraordinary qualities of every application. The advancements required, at length, a wide scope of levels and layers, from physical to information and application layers. One of these advanced ideas, when applied in a two-way transfer network with symmetrical recurrence division, was able to improve efficiency in clever meter correspondences on the lattice.

The force of brilliant and self-arranging gadgets, which are strongly connected in worldwide network frameworks, is the IoT, a prime example. IoT is commonly characterized as a genuine item that has generally scattered and restricted capacity and handling abilities, which is designed to work on interdependence, execution, and security of brilliant urban communities and their frameworks.

This program has made venture opportunities available because it is now not applicable to 'Savvy Cities,' which was one of Prime Minister Narendra Modi's primary mission promises in the course of his political decision in 2014.

Development is fundamental for making supportable urban communities. Economic improvement is a pressing issue that requires quick activities and changes with respect to legislatures, industry, and society in general. Learning is a significant part in these associations' quest for manageable advancement, which works on their manageability and empowers supportable execution. The present time is more significant than any other in recent memory to put resources into innovation to make urban areas strong across planned and useful measurements. Transportation and water supply are completely reliant upon energy. If there is no energy, administrations cannot work on any area what was just mentioned. This will affect individuals' lives, either directly or by implications. As a rule, the city requires an adequate stockpile of water, admittance to energy, a compelling and productive method of transportation, a feeling that everything is good, and a feeling of safety. To deal with all these, an intermingling of innovation, regulatory endeavors, and government approaches was required. The idea of a brilliant city is the one that considers these elements. The aim of this chapter is to sum up and propose a meaning of a shrewd city. The issues include an exhaustive report on savvy urban areas in India, zeroing in on highlights, approaches, current status, and future possibilities.

2.2 THE DEFINITION OF A SMART CITY

The expression 'brilliant city' is an inexactly characterized idea. Be that as it may, it is just characterized regarding the gathering of different assets, advances, and authoritative exercises for individuals' prosperity and maintainable turn of events. The expression 'savvy city' is likewise characterized as 'Metropolitan Intelligence' or 'Knowledge in Urban Life,' which empowers solace and security in all parts of individuals' lives. This definition or importance of a shrewd city might contrast from one spot to another, individual to individual, and government to government, which is contingent upon their methodologies and assets.

Brilliant urban communities, by and large, comprise different areas that require equivalent improvement with regard to ability, accommodation, and intelligence. The creators endeavored to make sense of brilliant city physiognomies using the 3-C (skill, comfort, and keenness) idea in this chapter. Savvy city drive has opened entryways for computerized reasoning exploration universally and laid out completely new innovation principles. The smart city, as imagined, is isolated into four layers: foundation, energy, transportation, and individuals' way of life. Gathering and sharing all the data more authentically across each of the four areas, as well as interfacing the organization layers, will work on metropolitan life and reinforce the structure holding the system together. Data and correspondence innovations (ICT) and man-made brainpower (artificial intelligence [AI]) are basic advanced instruments for the improvement of a smart city.

The idea of brilliant city is essentially founded on cost enhancement and worked on to meet expectations for everyday comforts, asset conservation, innovation mix, and quicker exchanges in all fields. It joins all parts of innovation to change a complicated framework into a computerized, modern, and less difficult lifestyle. In this unique circumstance, the expression 'innovation' alludes to the most famous and emerging fields, like AI and IoT.

2.3 INTERNET OF THINGS

An IoT system collects data from a large number of sensors. This data must be reduced in size so that it can be handled by existing networks, screened to ensure integrity and compliance with security policies, and processed and stored so that the eventual consumers can use it. A smart city must be instrumented, interconnected, and intelligent. This is accomplished by using IoT. By installing sensors (RFID, IR, GPS, laser scanners, and so on), connecting them to the Internet, and achieving intelligent recognition, tracking, monitoring, location, and management, it enables remote monitoring, control, and management of devices, as well as access to and analysis of real-time data.

2.4 CHALLENGES OF THE INTERNET OF THINGS

Although the IoT application domains and situations described before are very interesting in terms of providing technologies for smart everything, there are some challenges to implementing the IoT concept in terms of cost. It is expected that the technology will be of low cost and applicable to a wide range of objects. IoT is also confronted with a slew of other challenges, including the following:

- **Adaptability:** Because things team up in an open climate, the IoT has a bigger idea than the customary Internet of PCs. Therefore, fundamental usefulness, for example, correspondence and administration disclosure, should function admirably in both little and huge scope conditions. The IoT requires the improvement of new capacities and strategies to accomplish adaptability and proficiency.
- **Self-organizing:** Smart things ought not to be overseen similarly that PCs are, expecting clients to arrange and adjust them to explicit circumstances. Portable things, which are used rarely, should have the option to lay out associations all along and arrange and design themselves to suit their particular climate.

- **Volumes of information:** Some Web-of-Things application situations will include rare correspondence and social occasion data from sensor organizations, or from coordinated operations and huge scope organizations, which will gather enormous measures of information on focal organization hubs or servers. Huge information is the term used to portray this peculiarity, which requires various functional systems as well as new advances for putting away, handling, and the board.
- **Information understanding:** To help clients savvy things, it is important to decipher those that are not entirely settled by sensors as precisely as are really expected. To benefit from the dissimilar information that are developed, specialist co-ops should have the option to make a few generalizable inferences from the deciphered sensor information.
- **Interoperability:** The data handling and correspondence capacities of each sort of a brilliant article in the IoT differ. Different brilliant items would likewise be exposed to different circumstances, like energy accessibility and correspondence transfer speed necessities. Normal principles are expected to work with correspondence and participation among these articles: 126–148–139-World Scientific News 67(2) (2017).
- **Programmed discovery:** In unique conditions, proper administrations for things should be recognized naturally, which requires a suitable semantic method for portraying their usefulness. Intricacy of programming: To deal with the savvy protests and offer types of assistance to them, a more broad programming foundation on the organization and on foundation servers will be required. This is because of the way that the product frameworks in savvy objects, as conventional inserted frameworks, should work with negligible assets.
- **Security and security:** Other necessities would be significant in an IoT, notwithstanding the security and insurance parts of the Internet, for example, correspondence classification, the realness and dependability of correspondence accomplices, and message uprightness. In IoT, certain administrations should be obtained to or kept from speaking with different things, and deals, including shrewd items, should be shielded from contenders' intrusive eyes.
- **Adaptation to noncritical failure:** Objects in the Web of Things are substantially more powerful and portable than web PCs, and they change quickly and unexpectedly. Building a strong and dependable IoT would require overt repetitiveness at different levels as well as the capacity to naturally adjust to evolving conditions.
- **Power supply:** Because things ordinarily move and are not associated with a power supply, their knowledge should be controlled by an independent energy source. Albeit uninvolved RFID transponders don't need an outside power source, their usefulness and correspondence range are seriously restricted. Future low power processors and interchange units for implanted frameworks that can work with essentially less energy are expected. Energy protection is a thought in equipment and framework design, yet additionally in programming. Yet, it additionally applies to programming, for example,

the execution of convention stacks, where each and every transmission byte should legitimize its presence.
- **Remote correspondences:** Remote correspondence often occurs in areas with limited connectivity. Ensuring that IoT devices and AI systems work reliably under varying network conditions is a significant challenge.

The objective of savvy urban communities is to extend the abilities of current framework. In the case of a pandemic, these networks will depend vigorously on shrewd metropolitan needs to increase, generally speaking, versatility to such surprising occasions and protect their residents. As indicated by the United Nations, the worldwide population would have moved to 9.8 billion individuals by 2050. While our population density increases, a significant shift occurs, pulling individuals from provincial regions to metropolitan focuses. As indicated by the Population Reference Bureau, the reasons for this relocation of financial open door and personal satisfaction will move an expected 70% of the total population to urban communities when we contact 10 billion individuals. This implies that millions of additional individuals would live in similar urban communities as we do now, which are as of now overburdened in regions like power, transportation, and water treatment and circulation. We can likewise expect the ascent of a bigger worldwide working class, which will be very much situated to become rich buyers. Urban communities all through the world will require monstrous foundation and productivity upgrades to oblige this degree of population density.

2.5 SMART CITY CHARACTERISTICS

Sustainable practices in terms of technology usage, governing policy, economic viability, and so on are among the characteristics. Land area must be used efficiently in both planned and unplanned areas. Housing and other infrastructure facilities for all city residents could be expanded, which would help to improve the quality of life. Another critical factor is the development of service facilities that are within cycling or walking distance. This reduces the use of automobiles, lowering air pollution and resource consumption. Such characteristics will help the city's economic development.

Another significant feature is the creation of public utility facilities such as recreational areas, playgrounds, and parks in designated local housing communities. Intelligent transportation systems, particularly high-quality, low-cost public transportation, are now available.

Economic development activities such as industry, tourism, education, and sports facilities, among others, will benefit from friendly governance policies, including public-private partnerships.

1. **Security of the public:** Savvy urban communities are liable for something other than saving fuel and diminishing gridlock; they are additionally answerable for saving lives and fighting wrongdoing.
2. **Acknowledgment of patterns:** The main part of AI in savvy urban communities is resident city collaboration. It supports the review and examination of existing data to settle on better choices in the future while managing

a few massive changes in their activities. Many difficulties might emerge during the use of an unmistakable model because of the intricacy of urban areas. Choosing the best model for every city is basic to lessening the monetary gamble of huge forthright capital expenses.

AI arrangement procedures like Bayes Network (BN), Nave Bayesian (NB), J48, and Nearest Neighbor (NN) can be used to make precise forecasts information, especially temp and shady climate.

3. **Structures of brilliant urban communities:** The normal structures in SCs, as well as explicit designs and transformations of normal ones, are introduced in this part.
4. **Structures in like manner:** Since the origin of the SC worldview, different analysts have attempted to foster the most fitting design in view of ICT arrangements. In view of the different necessities and day-to-day circumstances in urban areas that have executed SC models, these executions have not kept a guideline and, therefore, they have various qualities. That is, there is nobody size-fits-all design for SCs; in any case, there are a few famous ideal models that address the issues of urban communities. The qualities of the layers responsible for securing, handling, and using information recognize these structures. Among them, the most well-known ones are (1) distributed computing, (2) mist registering, and (3) edge processing. These structures are integral and can be executed in a similar city; for explicit ventures or applications, see Longo et al. (2017). As discussed by Almeida et al. (2019), engineering level plan choices should be totally analyzed to survey their effect on the last framework.
5. **Design of distributed computing:** The National Institute of Standards and Technologies (NIST) characterizes distributed computing as follows:

   Distributed computing is a model for giving universal, helpful, on-request network admittance to a common pool of configurable figuring assets (e.g., networks, servers, capacity, applications, and administrations) that can be quickly provisioned and set with negligible administration exertion or collaboration free from administration providers. By default, this engineering is separated into layers, for example, Infrastructure-as-a-Service (IaaS) and Platform-as-a-Service (PaaS). Each layer consumes the administrations given by different layers while likewise offering its own types of assistance.

   As the engineering is partitioned into administrations, its groundworks are in data decentralization and calculation.

   Kaur and Maheshwari (2016) proposed a medical services structure in light of the introduced design in their undertakings that pre-owned this engineering. One of the studies made a structure for brilliant city security that uses distributed computing engineering to coordinate blockchain with IoT gadgets. Mazza et al. presented a versatile distributed computing system for depicting information streams and activities in shrewd urban communities. Khanna and Anand (2016) proposed an IoT and distributed computing engineering–based observed stopping framework.
6. **The engineer of fog computing:** Bruneo et al. (2019) displayed a clever adaptable use case that makes the use of the haze-processing worldview as

one of the tasks that pre-owned this design. Chen et al. (2018) developed a novel video exchange managing strategy based on fog computing engineering to provide real-time data handling and navigation. Santos et al. (2016) recommended a coordinated effort center and the leaders outline to trade configuration files among centers. Barik et al. (2018) developed an extraction of geographical information inquiry structure based on haze processing.

7. **Design of edge registering:** As an augmentation of the cloud figuring worldview, Bar-Magen Numhauser of CISCO instituted the expression 'mist processing.' This is achieved by working on the exhibition and obligations of the organization's end-hubs. Moreover, the distinctions depend on the establishment of the information stockpiling and handling gadgets in similar neighborhood network as the gadgets gaining the information, bringing about low-idleness correspondence between them. Due to these qualities, this design is used for applications requiring low inactivity, topographically scattered frameworks, or applications where it is unfeasible to send information to the cloud and sit tight for it to be handled.

Edge registering is a worldview that follows the reason that assuming information is developed at the organization's edge, which ought to likewise be more effective to deal with that information at the organization's edge. This is a comparative way to deal with mist figuring, then again, actually haze centers around the framework and edge on the things side. It ought to be noticed that the edge isn't generally the littlest sensor in an organization, yet rather the connection between the neighborhood organization and the cloud.

Concerning projects that pre-owned this design, Shi et al. (2020) made sense of why this worldview varies from the others referenced already and when carrying out it would be valuable.

Taleb et al. (2017) proposed a task that uses versatile edge figuring to work on clients' encounters with video real time in SCs. Wang et al. (2020) presented a framework for digital physical-social frameworks in light of cloud and edge engineering. In addition, some authors relate other themes to Cloud Computing in their studies; amongst them the Internet of things (IoT) and Big Data deserve highlighting; such technologies make use of Cloud Computing in their solutions (Srivastava & Kiran 2016).

The structures introduced in this part have been used in various ventures. These undertakings resolved the issue of SCs by giving different end-client arrangements. As recently expressed, every city is novel, so engineering should adjust with them. Besides, the structures talked about previously are not totally unrelated, so a similar city can involve various designs for various administrations.

Energy, versatility, structures, public administrations, and coordination are the other six parts that make a shrewd city. As per Water World, the reason for these exercises is to make the city more manageable and proficient, while likewise working on personal satisfaction. Savvy water is an expansive term that alludes to a comprehensive way to deal with this important asset and the framework that helps its inventory, treatment, and circulation. It's important that these organizations communicate with each other as we update and put resources into our water framework with more web-empowered advancements, and a plenty of information open up. Not exclusively will this empower the estimation of fundamental files like repository and groundwater supply, but also the emergency of framework alterations; however, it will likewise increase efficiencies across a few water-related disciplines. While taking a gander at recorded flood information joined with ongoing and estimated meteorological and precipitation information, Water World refers to the prescient capability of flood planning for instance.

Urban communities can improve and limit water waste and cost in appropriation by spotting anomalies in utilization designs for the two utilities and end clients. The high energy demand of a city's water treatment and conveyance networks is often misjudged, which implies that rising functional proficiency through significant information will at the same time decrease ozone harming substance discharges and lowers costs.

As per a UN report, India is expected to build more than 60 urban areas with a population density of 1,000,000 individuals and 6 megacities with a population density of more than 10 million individuals by 2030. This quick urbanization and population relocation to encompassing urban areas have placed a huge strain on the country's general foundation.

It is indispensable to achieve water security as it will support connecting the organic market hole. It's basic to fulfill this flood in need while limiting negative natural results. A shrewd city's supportability and metropolitan improvement ought to be considered two essential and fair components.

With full intent on defeating this test, the pandemic has animated examination and execution of new advanced innovations. The change of urban areas into shrewd water urban communities requires a complex course of digitalization of water. It is, be that as it may, the best way to offer great support to people in general, while additionally guaranteeing a more secure and more feasible future. Water networks are under gigantic strain because of increasing population density and worldwide environmental changes. As indicated by the United Nations, there were 7.6 billion individuals on the earth on March 1, 2018, and this number is supposed to increase to almost 10 billion by 2035, putting a strain on important water supply. Breaks and burst pipes are brought about by an old framework that is overburdened by a rapidly increasing population density, extreme precipitation, and flooding. These issues are as of now influencing numerous Indian urban areas, including Bangalore, Mumbai, Delhi, and segments of Haryana. These water misfortune and tainting influence an association, not simply by expanding functional expenses but also by neglecting to convey water 24 hours every day, except it likewise has a critical natural effect. The strategies to ending this water misfortune are being sought after with enthusiasm.

Considering these variables, brilliant city endeavors are at a junction. The water board frameworks that are savvy are imperative to the outcome of smart cities.

A brilliant water arrangement is a heap of items that gathers significant and noteworthy information from the water network progressively. Water utilities are producing the very measure of information in one day that they did in a year when brilliant water arrangements are introduced. The issue that utilities are presently confronting is that this information is held autonomously and for various reasons in various regions of the association.

At meeting room tables, AI is as of now a hot issue. Human insight will be intensified by AI, permitting savvy urban communities to flourish more than ever. Water utilities can assemble and associate ongoing information from their water organization to accomplish continuous information-driven business efficiencies by putting resources into AI advances. This data permits the water utility to all the more likely identified spills, pinpoint spill areas, confirm spill alerts continuously, kill misleading break cautions, further develop ongoing organization checking, right away answer organization or water quality issues, oversee strain to forestall explodes, and further develop client care to give water 24 hours every day, seven days per week.

To achieve 24 × 7 water supply and savvy city development, utilities and legislatures should use cutting-edge brilliant water innovations. Moreover, getting wastewater an exclusive treatment would give an extra and important wellspring of water for a city, and this water reuse is basic to diminishing strain on existing supply. Numerous people group all over the planet have currently really embraced innovation to moderate important assets. It is previous time for India to execute water protection measures inside and outside to hold water as much as possible and reuse water. Now it is the time to implement it second time!

As indicated by Vaidya (2018), a brilliant water arrangement is a heap of items that gathers significant and noteworthy information from the water network progressively. Water utilities are creating the very measure of information in one day, what they did over one year, once brilliant water arrangements are introduced. The issue that utilities are at present confronting is that this information is held freely and for various reasons in several regions of the association.

Brilliant water, as per Taddune (2018), is an all-encompassing technique to deal with this indispensable asset and the infrastructural frameworks that encompass its collection, treatment, and dissemination. It's important that these organizations communicate with each other as we update and put resources into our water framework with more web-empowered advances, and a plenty of information open up. Not exclusively will this empower the estimation of fundamental files like repository and groundwater supply, but also the emergency of foundation alterations; however, it will likewise increase efficiencies across a few water-related disciplines. While taking a gander at authentic flood information joined with continuous and anticipated meteorological and precipitation information, Water World refers to the prescient capability of flood planning for instance.

Sanjeev Sirsi (2021) concluded that IoT, man-made brainpower, and Big Data advancements can be used to effectively span holes in India's current water and wastewater framework.

As per Herrando (2021), urban areas, being the operational hubs of our reality, experience the very issues that we experience as individuals. The quality and execution of the water supply, as well as some other parts of its assortment, conveyance, or supply, can be in every way improved. The execution of water and energy in a smart city might experience the most troublesome errand of the ebb and flow digitalization. Computerized change isn't an objective all by itself. The benefits are self-evident; however, it additionally lays the preparation for sending the state-of-the-art-specialized arrangements. It will turn out to be dynamically hard to execute arrangements in the event that activities are not acted in a coordinated way.

Talari et al. (2017) expressed that because of the current development of computerized advancements, the savvy city is becoming more brilliant than before. Savvy urban communities comprise different kinds of electronic gear that are used by different applications, like cameras in an observing framework or sensors in a transportation framework. Moreover, the use of individual portable gear can be scattered. Therefore, considering the heterogeneous climate, different terms like article qualities, members, inspirations, and security strategies would be explored. The reference illustrated a portion of the vital qualities of homogeneous savvy urban communities in 2020. Brilliant residents, shrewd energy, savvy structures, brilliant versatility, and brilliant innovation are on the whole instances of savvy advancements.

2.6 SYSTEM OF WATER AND WEATHER

Climate frameworks use an assortment of sensors to give exact information like temperature, downpour, sun-powered illumination, and wind speed as well as to help with the proficiency of a shrewd city. Water circulation frameworks, similar to power appropriation frameworks, are basic parts of any brilliant city. Regular water dispersion techniques from the water source to the client premises are inadequate and wasteful, especially for diagnosing spills ready to go or different pieces of the framework. Subsequently, by putting sensors in essential areas all through the dispersion framework, it becomes adequately wise to distinguish any kind of shortcoming or other applications. A water source, like a lake or a stream, storage spaces, like supply, and conveyance organizations, for example, under-or over-the-ground pipelines, are generally parts of a water dispersion framework. The situation of sensors in an IoT-based climate can be utilized to recognize blames like spillage, water quality, repository water level, and the area of sensors in different pieces of water appropriation frameworks. Accordingly, the degree of water in a capacity is not entirely determined by putting an ultrasonic sensor on top of the tank and two strain transducers at the bottom. A glass anode for estimating water pH can be used to test the nature of water both before and after the capacity tank at vital areas. Spillage in pipelines can be identified using three unique sensors, vibration (using double pivot accelerometer), pressure (using piezo-resistive sensor), and sound (using ultrasonic sensor) monitoring. When a city gets excessively or too little downpour, large numbers of areas have genuine water issues. Easily available measures can use IoT to foster imaginative techniques to design and oversee water system, tackle extreme water use, further develop water protection, and all the more actual apportion of their scant assets, and address flooding and wastewater execution during a tempest, contingent upon the requirements of affected areas. In a district with no administration or plan, crude sewage will unavoidably be unloaded into nearby streams. IoT helps neighborhood legislatures in showing them how to oversee and design everything connected with water; for example, how to control a lot of crude sewage and how to be more proficient. Storm arrangement is fundamental. Urban areas can gather information for deciding the situation with water supply by using frameworks that screen the climate and control water capacity. Urban areas will actually want to lessen sewer flood and water defilement with the use of a reasonable technique. Moreover, flood episodes can be anticipated using a sensor framework that gathers information on the water level of groundwater sources and waterways. These framework sensors, which depend on ultrasonic reach locaters and are set close to streams to gauge changes in water level and foresee flood chances, use ultrasonic reach locaters to quantify changes in water level and anticipate flood chances. IoT empowers continuous improvement through TV whitespace stations made accessible by neighborhood telecom suppliers.

2.6.1 Role of Sustainable Water Management

Water access is a basic part of shrewd city foundation. Urban areas will encounter an immediate impact on their water use as they develop. It is essential to offer occupants with this entrance in a practical way by smoothing out water processes, exploring options in contrast to conventional procedures, and lessening waste and the energy use of these frameworks.

Internet of Things (IoT), AI, and Big Data are instances of advances that can be used to effectively span holes in India's current water and wastewater framework. Ahmedabad, for instance, has robotized its water treatment and circulation framework, with ongoing checking and upkeep alarms conveyed to the fitting power.

Considering that most of the nation is water-pushed, brilliant urban communities should guarantee that inhabitants have sufficient water and that it is effectively available. For instance, in the late spring of 2019, Chennai had intense water scarcity, with each of its reservoirs running vacant. By using information sharing and imaginative arrangements, neighborhood authorities are currently effectively running after taking on supportable water techniques.

There are different proactive activities that can well affect the current water circumstance, including steps to fortify the dissemination organization, desalination, wastewater treatment destinations, and water metering.

Investigating options to groundwater consumption would be a primary piece of upgrading water supply. Groundwater levels have been rapidly dropping lately, attributable to inadequately overseen extraction and impractical use. The water tables in urban areas like Ahmedabad, New Delhi, and Bangalore have dropped fundamentally.

Therefore, the appropriate wastewater treatment in brilliant urban areas over the nation has significantly developed. India currently has the ability to treat around 37% of the complete wastewater generated in the country consistently. Savvy urban areas can use a round way to reduce their dependence on freshwater by recycling water a few times. Wastewater treatment plants with proper advances can be laid out in the city to effectively treat civil waste. People and associations can likewise set up wastewater activities in their structures on a more modest or bigger scope relying upon their requirements. Singapore has had the option to meet 40% of its water interest by executing shrewd wastewater the board.

Admittance to cooling has additionally turned into a need in tropical nations like India. Cooling has turned into a standard part in all structures because of its expanded buying power. Central air applications that are harmless to the ecosystem have turned into a significant piece of savvy city advancement. Therefore, it's essential to take a gander at ways of saving a great deal of water and energy by using more productive HVAC frameworks.

2.6.2 Water Consumption

At any point, do you consider how that water got from the source to your spigot?

This requires a huge measure of energy, which is in some cases disregarded. Lessening the consumption of energy in running drinking water frameworks is vital as urban communities endeavor to become more astute and more maintainable.

By 2020, 56% of the total population were living in urban communities, putting extra demand on assets like clean water. The need to protect this significant asset is more earnest these days: by 2025, a big part of the total population will be living in water-pushed regions, as per the World Health Organization. Anyway, how might we exploit the IoT period's possibilities and use shrewd innovation to make our water consumption as proficient as expected?

Enter … insightful water! The buzz around savvy urban areas is all over, yet before brilliant urban areas can turn into a reality, we should initially make water shrewd. Building water frameworks, some of which have been set up for more than a century, should be refreshed and improved with IoT advancements to come on the web and associate with different components of the framework and cities to achieve brilliant water.

In any case, what precisely are savvy water frameworks, and how would they work? Savvy water frameworks, similar to brilliant energy frameworks, catch constant information using IoT-empowered sensors.

This considers the improvement of water offices by recognizing breaks and observing how water is disseminated across the organization, permitting people to settle on better water the board choices. These savvy sensors, for instance, can recognize spills in water pipes and expeditiously inform engineers, permitting them to mediate and restrict the impacts. This is very helpful, as it is assessed that 3.3 billion liters of water are lost consistently in England and Wales due to infrastructural breaks.

2.6.3 Digitalization All through the Water Cycle

Siemens (2021) made the Digital Enterprise, a complete scope of programming and computerization advances, to assist the water business with taking advantage of digitalization. Handling all suitable information in a thorough information model permits all frameworks to be used to their maximum capacity. Therefore, a computerized twin is made, which is a precise, virtual model of the plant that is exceptional all of the time.

2.6.4 Benefits for Greenfield and Brownfield Plants

The intelligent integration of data from various sources, such as sensors, water meters, and meteorological data, is providing new opportunities in industry, agriculture, and municipal utilities to use water more efficiently – and thus more sustainably. A prerequisite is end-to-end system engineering networking, from commissioning to operation, maintenance, and continuous process optimization based on a data platform – a blending of the virtual and real worlds.

In practice, improved data quality and availability shorten project execution time. In the virtual world, on the other hand, simulation and modeling save incalculable costs during the development process as well as ongoing operating costs.

2.6.5 Recent Water Management Trends

- **Request-driven distribution:** A water appropriation framework in view of interest based conveyance can guarantee that the asset is dispersed reasonably. Unreasonable tension is the essential driver of water spills in the circulation framework. A definite pressure framework is strictly required to limit water squander.
- **Water metering and governance:** People usually underrate the volume of water they consume on a regular basis. Brilliant meters, rather than standard meters, are actually required to give constant information and an accurate comprehension of water use patterns in shrewd urban communities. Residents will be made to use water mindfully due to these meters. They can be compensated or rebuffed relying upon the idea of their consumption. During dry spells and other outrageous climate, information from savvy water meters can be acquired to restrict water use to its necessary level and control water use for different exercises such as vehicle washing and cultivating. These actions are similarly basic for accomplishing long haul advanced change in brilliant city water foundation.
- **Region cooling:** This has appeared a reasonable answer for cooling a number of districts in a financially savvy and harmless way to the ecosystem. At their GIFT downtown area, Rajkot was the first savvy city quite before to use locale cooling system. This has allowed them to monitor a huge volume of water and energy. Different urban communities, including as Amravati and Thane, are attempting to introduce locale cooling as well.
- **Coordinated approach:** A system for brilliant urban areas can be made by the board through adopting an incorporated strategy to economically maintain the use of water and reduce wastewater. It works with better water asset and cycle coordination by the board for an assortment of utilizations.
- **Advancements:** The execution of essential water volume and reducing wastewater is similarly reliant upon the approach and attitude of individuals and partners in shrewd urban communities, notwithstanding interests in framework and water processes. As urban communities welcome maintainability and new advances, it is similarly imperative to raise information and refinement about the significance of sensible utilization of existing water supply. We've perceived how wastewater treatment innovation supports the location of the infection, as well as how programmed limit control frameworks help in the anticipation of its spread. Nevertheless, it has become clear that significant work still needs to be performed around here, and that, most importantly, an innovative framework should be considered; and, needless to mention, for the turn of events and implementation in any field, public cooperation is required.

2.7 APPLICATIONS

2.7.1 Managing Water Wastage

According to an India Today article, around 40% of piped water in India is lost due to leakage. According to a US EPA research, a typical family can waste 180 gallons of water per week, or 9400 gallons per year, due to leaks in the home, which is fair enough for washing more than 300 loads of laundry.

We waste a lot of water due to leaks, burst pipes, and other issues, and AI and IoT can help us get rid of facing these problems. Water wastage can be reduced by using AI to evaluate real-time water loss and automate pipes to cut off if there is a leak. AI can detect storage tank leaks and assist in their repair before it's too late. IoT-connected devices can better interact and integrate multiple systems throughout a city or a locality.

The most important variables for making an informed decision were highlighted as

1. Motives for starting a wastewater management program,
2. Policies and regulations,
3. Technology and financial resources that are both available,
4. Scale of intervention,
5. Institutional framework and management strategy,
6. The public's opinion,
7. Various stages of deployment,
8. A framework for a collaborative approach.

2.7.2 Wastewater Treatment

AI technique can be used to minimize pollutants in the water, resulting in less water contamination and resolving the scarcity of clean water. Because AI is based on optics, it may be used to detect the number and composition of dangerous substances, which can improve the efficiency of waste management systems. Water quality can be continuously monitored, and real-time data on the quality can be obtained using machine learning and big data.

The process of purifying sewage that enters from diverse sources is known as wastewater treatment. The general treatment procedure utilizes bacteria to 'digest' the sewage and separate it into clean water and solid sludge once the primary solids come out of it. This treatment method necessitates the use of energy, oxygen, and chemicals.

We employed sensors to determine how the plant's requirements for oxygen, energy, and chemicals were impacted. Then we could generate and plan activities, as well as how to adapt the usage of these resources to be the most efficient and waste-free.

The plant's overall electricity usage was reduced by 13.5% as a result of the system's utilization. The facility also makes a better use of its resources, reducing the amount of chemicals needed to remove phosphorus from the water by 14% and reducing sludge generation by 17%.

2.7.3 Using IoT to Conserve Resources

At different periods of the day or night, different volumes of water arrive. Furthermore, the microorganisms that break down the wastewater respond differently in cold conditions. The cost of electricity varies depending on the time of day or night. However, utilizing IoT to enhance diverse systems has the potential to transform many facets of daily life, including healthcare, transportation, insurance, and a variety of other businesses.

2.7.4 Digital Water

With data analytics, regression models, and algorithms, AI can make water management easier. These cutting-edge technologies aid in the development of efficient water networks and systems. AI can be used to design water treatment plants and determine the state of water supply. Water managers and government agencies can utilize AI to create a smart water system that can provide efficient water infrastructure and adapt to changing conditions. These technologies will be cost-effective and long-lasting, allowing for the optimization of all water management solutions as well as the prediction of possible damage.

2.7.5 Smart Irrigation

Agriculture is the largest consumer of water, and many regions rely heavily on groundwater for irrigation. Smart irrigation will use AI systems to reduce water consumption and waste while optimizing water supply as well. AI systems can monitor groundwater levels and estimate agricultural needs in order to balance water usage by directing sprinkler systems.

More advanced precision-based AI systems can forecast weather, climate, and humidity, allowing for improved agricultural management. Using AI sensors, smart farms will be able to eliminate leaks and analyze the soil to identify the state of the plants and their water requirements.

The use of AI in water resource management has shown its ability to distribute cost-effective water and enhance water conservation systems.

2.7.6 Water Management with Intelligence

Water protection is a significant component of safeguarding water, particularly in metropolitan regions where it is challenging to monitor water consumption. This issue can be cured by implementing IoT, which will monitor how much water individuals consume on a particular day. We can use the information and the weather patterns of the day to decide how much water people in a given city should use. This will undoubtedly help the water experts in a more actual way, dealing with the progression of water.

2.7.7 Innovation

Water is essential for individual's existence on this planet. Nonetheless, as of late, we've underestimated the asset. Researchers are looking for better, stronger ways to deal with the execution of water asset considering the surprising expansion in the event of dry seasons and other water shortage emergencies in many areas of the planet. However, the need to effectively oversee water and its assets hasn't changed over the long haul, researchers have tracked down that developing and supporting existing water of the board frameworks requires an alternative system.

Researchers are at present laying out economical water supplies for individuals from one side of the planet to the other by involving AI in water asset of the board.

2.7.8 Man-Made Intelligence in Water Resource Management

Man-made intelligence, the sacred goal of innovation, is projected to introduce the 'advanced water' pattern. Simulated intelligence can work with huge volumes of informational indexes endlessly. The innovation likewise learns as it goes, improving with each new group of information it experiences. Due to its flexibility, AI is the ideal answer for managing a continually changing element like water.

In basic terms, computerized water will be overseen by information investigation, relapse models, and probabilistic calculations. With such tech-driven patterns empowering the board of water and government associations to foster effective water frameworks, the world's water networks are in for a ton of progress sooner rather than later.

2.7.9 Building Effective Water Frameworks

The bodies responsible for water execution have two significant positions in front of them. Because of the asset's capriciousness, the main issue is to track down new wellsprings of water, and the second is to economically oversee ebb and flow water stocks and framework.

Specialists are trying different things with calculations and programs to make water get established that can not only give flow data on the condition of an asset but also additionally help in the improvement of future models. Water tasks are planned progressively and insightfully utilizing programming with sensors and brain organizations. Expectation models for future water use, automated sensors in water treatment plants, and square chains for monetary exchange handling can be in every way made feasible with the right AI devices.

2.7.10 Advancing Current Water Resources with AI

As per UN gauges, around 1.2 billion individuals, or one-fifth of the worldwide population, live in water-scant spots, with further 500 million nearly doing as such. Another 1.6 billion individuals are confronting monetary water deficiencies because of the absence of foundation that permits them to get to perfect, consumable water.

The primary focal point of such insights is on the choice of augmenting present water supplies. Computer-based intelligence's dynamic capacities can be utilized to accomplish this advancement and mechanization. Government associations and water divisions can utilize AI-driven control to detect continuous water scarcity and misuse, assemble and execute broad dissemination organizations, and put forth a monetary objective of augmenting pay.

A group of scientists from the Key Laboratory of Eco-hydrology of Inland River Basin, Northwest Institute of Eco-Environment and Resources, Chinese Academy of Sciences, and an Australian partner attempted a progression of trials to decide groundwater save levels. To conjecture groundwater saves, they used vector relapse and other AI calculations.

Water frameworks over the world could benefit from half-breed man-made brainpower models. To add values to such a precious asset, researchers and experts should come up at a large scale.

2.7.11 IoT Architecture

The accompanying circumstances should be met to build a standard water of the board model and Figure 2.1 depicts water model:

REQ 1: Remote administration of actual components and activity of central units; recognizable proof of assets in the water organization, depiction of tasks and conditions over the organization ought to be in every way covered by the framework.

REQ 2: It ought to be viable with different applications like geographic data frameworks and data sets giving information on soils, weather conditions figures, the climate, cultivating, etc.

REQ 3: For the mix of different frameworks, it ought to have an adaptable and expandable design. It should make open points of interaction among correspondence and cycle control levels, as well as join IoT frameworks for direct admittance to individual water level execution gadgets, to achieve this.

REQ 4: It ought to have the option to work with verifiable frameworks and control current hardware. Water executives frameworks currently implemented in provincial regions comprise various interconnected and basic gadgets that should be overseen utilizing heritage strategies. They join correspondence highlights, information models, and conventions that are subject to the producer's restrictive innovation.

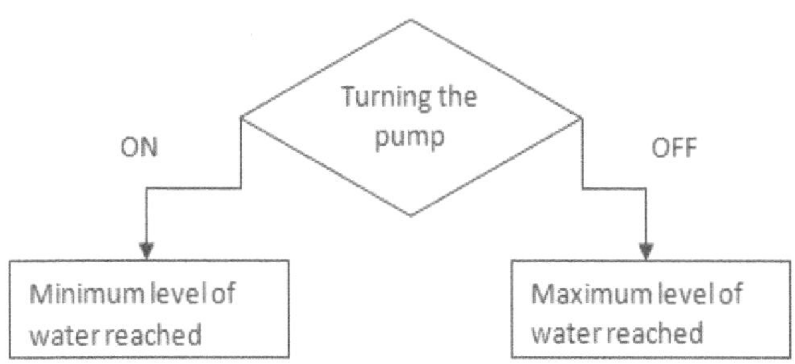

FIGURE 2.1 Water model.

A subsystem with four siphons and a tank is demonstrated in the situation. Water is drawn from a water treatment plant through the siphons. To control the genuine framework, the framework uses the accompanying principles:

1. Toward the beginning of the interaction, all siphons are switched off.
2. After a foreordained period, it checks the water level in the tank.
3. At the point when the tank hits the most reduced level, it initiates all siphons.
4. At the point when the tank arrives at the more elevated level, all siphons are switched off.
5. Whenever the volume of water going into the tank surpasses the volume of water streaming out of the tank, the siphon is switched off for quite a while.
6. At the point when the volume of water going into the tank is not exactly the volume of water streaming out of the tank, the siphon is switched off for quite a while.

The accompanying parts make up a customary water supply framework:

1. Raw water source: Water found in the climate.
2. Raw water siphons: An assortment of hardware and designs for gathering water from waterways and lakes.
3. Raw water transmissions: Transportation crude water to the water treatment offices.
4. Water treatment station: Infrastructure for water filtration was put near conclusive conveyance areas to get a good deal on siphoning and control contamination after treatment.
5. Water capacity tank: A water tank that is situated on top of a construction at a height that allows the water dissemination framework to be compressed.
6. Water dissemination framework: An assortment of interconnected lines and valves that appropriate water to end-clients.

Figure 2.2 depicts flow of methodology and Figure 2.3 depicts the proposed architecture.

The water management system is divided into three layers by the architecture.

2.7.12 Bottom Layer

This layer contains all components of the water supply system, such as raw water sources, raw water pumps, raw water transmission pipes, water treatment stations, water storage tanks, and the water distribution system. Water meters, IoT devices for measuring pressure in pipes and volume in a water storage tank, and actuators are also included. It connects a network of sensors and actuators to a gateway.

2.7.13 The Middle Layer

This provides storage, analysis, processing, and orchestration services for communication between the bottom and top levels of IoT infrastructure, as well as between sensors and actuators.

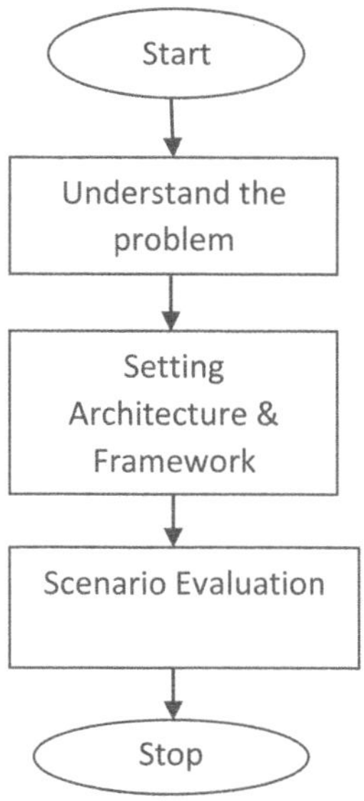

FIGURE 2.2 Flowchart of methodology.

2.7.14 Top Layer

The design enables for the creation of a variety of applications, such as performance indicator dashboards and web/mobile applications for water supply system administration.

2.7.15 Analysis and Discussion

Water management in real time is resulting in significant cost reductions. Last year, Hitachi Consulting was tasked by South West Water in the United Kingdom to optimize its water operations by integrating IoT solutions into the utility's network, allowing it to manage its operations more effectively with near real-time data and analytics. Hitachi is assisting the organization in managing the reliability of its network and improving the

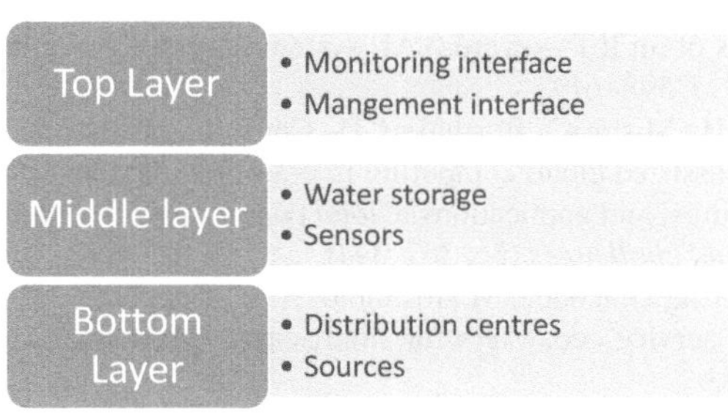

FIGURE 2.3 Proposed architecture.

service it provides to customers by utilizing cloud technologies. These advantages are also given to customers. Consumers can check their water usage remotely via applications thanks to smart water technologies. It can alert them to the presence of a leak, allowing them to address the issue much more quickly and avoid wasting water. Smart water gadgets are becoming increasingly popular, allowing users to monitor their own water consumption and make decisions based on the information they see, allowing them to use water more sustainably while also saving money.

This can be summed up as the desire to expedite technical innovation, which is exacerbated by an increasingly global environment. Citizens want government agencies to provide increasingly innovative services. These, in turn, must be nimbler than ever before.

2.8 CONCLUSION

Since people strive to make the most of the innovative abilities at our disposal in order to become more practical, it is critical that we do not overlook the asset that, all things considered, we cannot exist without. Water framework redesigns that are astute will not only help deal with the rising interest caused by mass urbanization, but they will also help reduce water waste and make our use more efficient.

We've perceived how wastewater treatment innovation helps with the location of the infection, as well as how programmed limit control frameworks help in the counteraction of its spread. Notwithstanding, it has become clear that impressive work still needs to be done around here, and that, most importantly, an innovative foundation that takes into account the turn of events and sending of coordinated arrangements in any field of public help is required. The significance of creating innovations, the IoT, and prescient examination with regard to smart cities has been inspected exhaustively. An assortment of innovations for smart water of the board framework is additionally tended to. An exploration of a few IoT-based cloud stages was directed. An idea for a shrewd water of the board framework in light of the IoT that is savvy has been introduced. The upside of the framework is to furnish the sufficient water supply with great quality water to each house, industry, and others. The proposed model can be executed as a piece of the shrewd city.

REFERENCES

Almeida, A., Mulero, R., Rametta, P., Urošević, V., Andrić, M., & Patrono, L. (2019). A critical analysis of an IoT—aware AAL system for elderly monitoring. *Future Gener. Comput. Syst.*, 97, 598–619.

Barik, R. K., Dubey, H., Misra, C., Borthakur, D., Constant, N., Sasane, S. A., ... Mankodiya, K. (2018). Fog assisted cloud computing in era of big data and internet-of-things: systems, architectures, and applications. *Cloud computing for optimization: foundations, applications, and challenges* (pp. 367–394).

Bruneo, D., Distefano, S., Giacobbe, M., Minnolo, A. L., Longo, F., Merlino, G., ... Tapas, N. (2019). An IoT service ecosystem for smart cities: The# smartme project· *Internet of Things*, 5, 12–33.

Chen, Y., & Han, D. (2018). Water quality monitoring in smart city: A pilot project. *Autom. Constr.*, 89, 307–316.

Herrando, S. (2021). *Smart water cities: Water management in the cities of the future*. Europe: IDRICA.

Kaur, M. J., & Maheshwari, P. (2016, March). Building smart cities applications using IoT and cloud-based architectures. In 2016 international conference on industrial informatics and computer systems (CIICS) (pp. 1–5). IEEE.

Khanna, A., & Anand, R. (2016, January). IoT based smart parking system. In 2016 international conference on internet of things and applications (IOTA) (pp. 266–270). IEEE.

Longo, F., Bruneo, D., Distefano, S., Merlino, G., & Puliafito, A. (2017). Stack4Things: a sensing-and-actuation-as-a-service framework for IoT and cloud integration. *Ann. Telecommun.*, 72, 53–70.

Mazza, D., Tarchi, D., & Corazza, G. E. (2017). A unified urban mobile cloud computing offloading mechanism for smart cities. IEEE Communications Magazine, 55(3), 30–37.

Numhauser, B. M. (2013). Fog Computing Introduction to a New Cloud Evolution. Escrituras silenciadas" paisaje como historiografía. Spain: University of Alcala (pp. 111–126). ISBN 978-84-15595-84-7.

Santos, J., Rodrigues, J. J., Casal, J., Saleem, K., & Denisov, V. (2016). Intelligent personal assistants based on internet of things approaches. *IEEE Syst. J.*, 12(2), 1793–1802.

Shagov, N. S., Mamedova, N. A., & Urintsov, A. I. (2021, January). The Construction of the Graph Model and Objective Function for the Cloud-fog-edge-user [CFEU] Hybrid System. In 2021 IEEE Conference of Russian Young Researchers in Electrical and Electronic Engineering (ElConRus) (pp. 1946–1950). IEEE.

Shi, Q., Dong, B., He, T., Sun, Z., Zhu, J., Zhang, Z., & Lee, C. (2020). Progress in wearable electronics/photonics—Moving toward the era of artificial intelligence and internet of things. *InfoMat*, 2(6), 1131–1162.

Siemens. (2021). The digital enterprise portfolio for the water and waste water industry. Germany: Market Specific-Solutions.

Sirsi, S. (2021). Sustainable water management for smart cities. India: Construction Times.

Srivastava, A., & Kiran, P. (2016). Transforming Customer Experience in E-Commerce Market Place: A Smact Approach. *Indian J. Sci. Technol.*, 9, 32.

Taddune, G. (2018). Water wise: The role of water management in the smart city revolution. Texas: Environment+Energy Leader.

Talari, S., Shafie-Khah, M., Siano, P., Loia, V., Tommasetti, A., & Catalão, J. P. (2017). A review of smart cities based on the internet of things concept. *Energies*, 10(4), 421.

Taleb, T., Samdanis, K., Mada, B., Flinck, H., Dutta, S., & Sabella, D. (2017). On multi-access edge computing: A survey of the emerging 5G network edge cloud architecture and orchestration. *IEEE Commun. Surv. Tutor.*, 19(3), 1657–1681.

Vaidya, A. (2018). Effective water management – Indispensable for smart cities. India: BW Smart Cities.

Wang, Q., Zhu, X., Ni, Y., Gu, L., & Zhu, H. (2020). Blockchain for the IoT and industrial IoT: A review. *Internet of Things*, 10, 100081.

3 Introduction to Digital Transformation

Mooon Paiithannkar[1] *and Jacob Alexander*[2]
[1]Sinhgad Technical Education Society, Pune, India
[2]Resources 4 Solutions Inc., Delta, British Columbia, Canada

3.1 INTRODUCTION

Every organization now faces a choice between innovation and extinction. Digital transformation is all about reinventing business. It is simply not about IT or technology but to redefine the whole business strategy and company culture. To succeed in digital transformation, organizations must not consider technology just a support function but rather a strategic competency.

It is actually about creating new business model on back end of technology. When we consider "Amazon.com", it is not a mere store, rather it has entirely revolutionized the entire concept of shopping. Organizations have become more imperative and adaptive to take advantage of the ways the digital world has radically changed consumer behaviour and expectations of

- customers
- channels
- competitors

which are all in the digital world. According to a recent survey, mobile devices are those in which most numbers of minutes are seen dominating online-actions such as Search, Reading, Browsing and Shopping globally.

3.1.1 History of Digital Transformation

In the 1980s and 1990s, when businesses had first started, spreadsheets, word processors, and other rudimentary types of technologies were used, which are very basic today. Over the years, particularly in the 1990s, there was a big revolution of enterprise software which went beyond spreadsheets and word processing that allowed company to track order, information of finances and sales information. It really evolved into, which is now known as enterprise resource planning (ERP). It started off as material resource planning. In many cases, manufacturing organizations were just trying to track their inventories and manufacturing operations which then evolved into ERP. Many organizations started to implement more robust accounting and financial software. So back in the 1980s and 1990s, one could really trace the evolution of digital transformation.

 DOI: 10.1201/9781003225317-3

3.1.2 Definition of Digital Transformation

It is the use of technology that helps improve business. That could be ERP software. It could be customer relationship management software to help manage the sales force. It could be financial and accounting software, which is a technology to automate the shop floor. If it's a manufacturing organization, it could be robotics, or Internet of Things, or Blockchain, or artificial intelligence. Digital transformation is one sort of technology that enables your business to be more productive, efficient, and effective. It is said that Digital Transformation = Business transformation.

The secret is that digital transformation really doesn't have a lot to do with the digital part of the term, but it has a lot more to do with transformation. Even though new technology has driven out some of these digital transformations, it is really at the end of the day more of a business transformation that organizations are trying to accomplish. Generally, they are trying to improve their efficiency and provide better customer experiences to their customers. They are trying to be more effective as an organization and make the jobs easier for their employers. So really it's driven by the business needs and it just so happens that technology is a way to make those changes in which the key word is transformation that is the operative word within digital transformation. We can actually scratch digital and say it a business transformation. Digital is likely to be a part of that business transformation, but there is a lot more to it than just the technology itself. There are also process improvements, process management, and process reengineering that need to happen, and there's also the organizational change or the people side of the equation that's going to change. So this is often why business transformation is perhaps a more descriptive term in defining what exactly digital transformation helps improve the business or the organization.

3.2 CONCEPTS AND IDEAS

It is a strategy for transforming the way the manufactures operate their business. It starts with admitting that there's a drag. The problem is that it doesn't operate on real-time data operations.

To solve any problem, we need to ensure that it's connected, integrated, integral, and at current state.

Digital transformation allows us to take things and make them virtual and transform them into things that happen in virtual world. As we see the use of smart phones to do the banking – internet banking without going to the bank.

This process is of virtualization on how things are taken from the real world and put into the virtual world. To look into the transformation part, the process is having huge impacts and that's really interesting. This word transformation is actually an important one in economics, sociology. It was used only once before to describe processes in our society, which is useful in informing about happenings.

The idea had come from the pre-industrial society that if there were ways in an area, there might be tons of changes, wars, new kings, queens all the time, which we observed in the history, but at the same time it was actually very stable. People's lives were similar for hundreds of years that the lives of grandchildren looked quite a lot like the lives of grandparents, where they lived the same way, they worked the same way, they found life partners the same way, and they practise their religion in the same way. Very little changes were there in terms of the lives of people right until something happened, but this is the first time that people have started using this word transformation and applying it to the society, which brought Industrial Revolution into existence. That's why a gentleman, Karl Polanyi, wrote a book called "The Great Transformation" that talks about the fundamental transformation of everything in our society held a hundred years ago and a little less than a hundred years before. We see the nature of work changed the articles that we used in our everyday lives. It changed the appearance of the mass manufacturing, the way printing press appeared. We had newspapers, our political system changed and interestingly for a long period of time, there was no change, and suddenly there were fundamental changes in every dimension of society. People's lives were completely different from one point to a different.

Example:

Grandparents could not recognize the life of grandchildren, which was the first big transformation. Similar kinds of things happen again, and everyone is in middle of it and is experiencing it, and it's a partial transformation that is complete. A new world is created, even if they look familiar, it is not real. If we take the example of a car, during the grandparents' time it was different from that of now. So much is electric and all that's interesting where a platform for software is seen. Once car gets software updates it changes with new capabilities. It can do new things that become semi-autonomous. This whole change is the digital transformation, which means a fundamental transformation of the basic way their industries work.

We are going through an Industry mutation-

Its not only transformation. It's a mutation.

Sebastien Bazin
CEO, Accord Hotels

Here he says how they are going through an industry mutation. It's not transformation. It's a mutation he's trying to highlight that what has happened is so fundamental that he doesn't even know how to do business any more what actually happened in this industry's "Booking.com" – where a company came and inserted themselves between the hotel chain and its customers and made business by 30% of the revenue which is exactly equal to the profit.

Another transformation of idea – we see – is in the newspaper reading, where we used to get newspapers dropped at our doorsteps every morning, but, due to digital transformation from the traditional one, now we can read it digitally instead of receiving it physically. It isn't just about business or government or big organizations or small organizations. It's also about one's life as an individual, which has also equally changed lots of things that are done traditionally. In companies like "AMAZON" and "APPLE", we always see something new happening in these – may it be the size and the kind of power. They always deal very successfully in the digital world. We know, "AMAZON" is bigger than the rest of 12 global retailing companies combined, and through their service, we can order any stuff from home at any time without going to anywhere physically.

Similarly, if we take the example of "NEWSPAPER", we can see a drastic transformation of business approaches in the world. We see news all the time on our handset devices which keep updating us constantly. You can check for updated news at any point of time.

The entire system of communication with friends has changed. The way of searching soulmates online has changed, for example, "Shaadi.com", "Jeevansathi.com", "Bharat Matrimonials.com".

We can see a structural new transformation taking place. We work in tandem with each other around these digital devices and are able to live in a different way.

3.3 AREAS OF DIGITAL TRANSFORMATION

Digital transformation revolves around these five areas of transformation:

1. Leadership transformation
2. Omni-experience transformation
3. Information transformation
4. Operating model transformation
5. Work-source transformation

3.3.1 Leadership Transformation

We are going through a process of digital transformation(see Figure 3.1), and the challenge is to figure out the kind of leaders we need to manage the process in a way that we get the advantage from it and we avoid the biggest dangers of it.

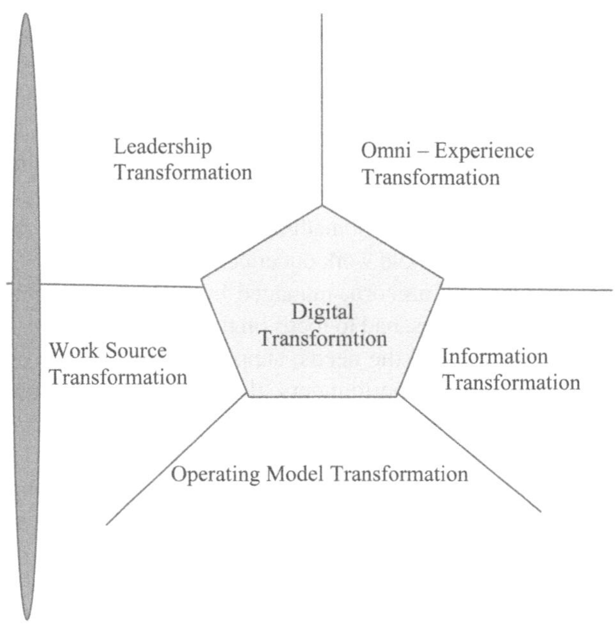

FIGURE 3.1 Areas of digital transformation.

A lot of technologies that we hear about are all kinds of potential horror stories as well as all kinds of amazing applications and all kinds of leaders we need to ensure that process works to transform the organization. Our economies transform means our lives as individual continue to be transformed by these technologies. If this technology comes up, it brings huge changes in organizations.

Are we prepared in terms of having the kind of leaders the people in place, in organization, in our government that can deal with the question that comes up and makes sure that these technologies are implemented in a way that provides the most benefit and the least damage in spending billions of money developing new technology on almost nothing developing new leaders. We need to really think very clearly about what these new leaders need.

> **Leadership competencies are leadership skills and behaviours that contribute to superior performance.**
>
> **Society for Human Resource Management**

Leadership competencies are the real idea that the leaders should be able to do in order to be successful in leading the organization and the people that they are leading.

We refer to technological competencies, which is understanding the technologies that one organization needs to deal with. Leaders are required in every organization that can deal with such technologies like Blockchain, AI and robotics. These technologies are very important and we need to get people up to spread in various events.

Let us consider the following examples:

TEST – A world where experimentation is much more important than it had traditionally been. Because we have a lot of unknown subjects in an organization, we must have these new technologies to face a number of challenges and lots of uncertainties; now it becomes more important for an organization to practise some trend of experimentation to enable itself to build teams. Cultures in an organization help people to get aware of the world, which is run through quick cheap experiments to come up with solutions. These are not mere experiments in a lab; rather they are to try something in a way that allows them to learn whether the experiment would work out effectively.

SHOPIFY – A Canadian unicorn, managed by a Canadian company worth more than a billion dollars, had to focus on the organization's strategic innovation, which gave rise to the needs, supply, and motivation of its people. Due to the young age of the founders, the nature of the organization and how they had to motivate their employees was very different, requiring a whole different way of doing that.

BENTLEY – The challenge for this company was not with young employers but with old employees in the company, with returnees. Employees were no longer retiring as they stayed healthy, fit and perfectly able to contribute to the companies.

A diversity of employees is seen working around with so many generations working together. So as a leader, challenge is how to deal and manage this diversity and leverage this kind of diversity. Most importantly, the ethical challenge that exists around these new technologies and organizations, when it comes about privacy, can collect lots of information on people and use it, channelize it in the right way, as law and ethics both prevail based on company's position and privacy based on the data that is collected to deal and protected. These new technologies are exponential, so the impacts of what we do with these new technologies and how to put them efficiently and effectively are very crucial. In traditional kinds of decision, the impact falls on the technological adoption. In AI systems while implementing and making decisions, the ethics and morals are embedded in the AI system. As it impacts lives deeply, so it impacts organization's work cultures. This can help leaders to manage and make the possible positive outcomes.

3.3.2 Omni-Experience Transformation

In this omni-channel world, it is critical to give customers "what", "when" and "where" they want. This created pressure on retailers and consumer products companies to ensure that the promise they make to the customer is satisfied by their supply chains. In future, companies will leverage innovations technology, new delivery methods and differentiating business practices to enable convenient prediction and cost-effective deliveries to their customers.

Definition – Omni-experience transformation means to interact across multiple channels as part of our seamless customer journey.

Example: Ishat clicks on a Facebook advertisement (see Figure 3.2) for a pair of cotton masks and is taken to the company's website. His size is not in stock online. So he interacts with an agent through online chat to see when they'll be back in stock. The agent tells him that his size is available in his nearest store, the agent reserves the pair of masks, and when Ishat goes to collect them from the store next day, Ishat had four different touch points with the company.

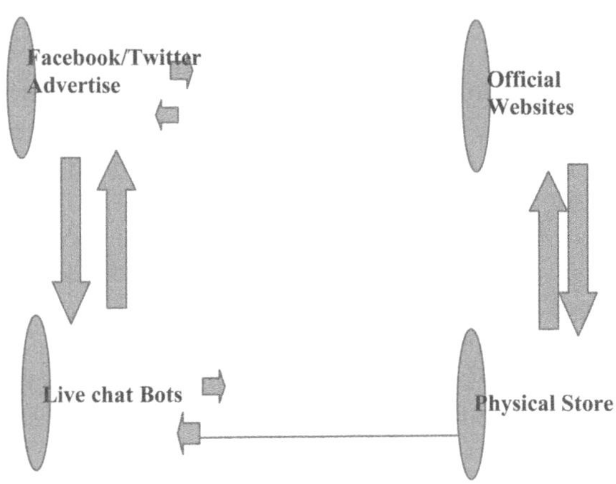

FIGURE 3.2 Diagram of the omni-experience transformation.

3.3.3 Information Transformation

Definition – The global businesses are changing rapidly. They are more inherently complex, especially organizations that are data centre, which are often chaotic, where 100s of servers and storage units, with multiple data bases and dozens of operating systems need to work together seamlessly to satisfy 24/7 users demand and the business process application requirements.

Such problems are faced by consumer goods manufactures, which come as no-surprise applications of artificial intelligence. Such machine learning and big data require tons of information in order to get outstanding results. A digital transformation is taking its shape towards a holistic view of their infrastructure and secure utility style computing model through an information transformation programme.

Digital business transformation implementation is the ultimate challenge for managers and executives. To survive, companies must go digital and adopt the latest technology. Digital transformation and information transformation impact not only the industry structures and strategic positioning but they also do it in different levels within an organization.

Information transformation is important for both growth and survival of any business. Digital disruption is a reality nearly every organization faces. Therefore, it is imperative that business leaders constantly challenge their companies and employees to ensure that information transformation unlocks productivity, gains a significant competitive advantage, all while delivering an exceptional customer's experiences.

According to economist – **Mr. Robert Gordon** – there are three industrial revolutions.

1950–1830 – Rise of steam engines locomotives of the railways, and the telegraph.

1870–1900 – Rise of electricity. The internal combustion engine indoor plumbing and water communications. Chemical plastics and a whole host of things that came about in that era.

The third industrial revolution (1900–2000) is the era of computers, web smart phones.

In the first two revolutions, India was not an independent country so was not able to leverage those revolutions, but in the third Industrial Revolution of Technology, information technology is happening at a time when the country is free and therefore it is uniquely used for development.

Here Moore's law has become so important in building power. The computer is becoming cheaper incessantly and more powerful day by day as smart phones which are equal to power of a computer which was used in the space programmes. Things are becoming more powerful and accessible to all, which is the rise of ubiquitous computing. An era of inexhaustible amounts of computing storage and space on the cloud which can be accessed through various Dropbox, Google Drive network, and this India saw the avalanche of that with mobile phones.

The best example that one can state is the Aadhar card enrolments/Unique Identification Authority of India is one such example of how the modern technology is being used to leapfrog. This was mainly to acknowledge the one who doesn't have any existence by people and their state due to lack of acknowledgement of their identity.

A biometric system was used to collect information of every individual for enrolments with uniqueness and no duplication.

This was possible only because of advancement in technologies. Internet has become pervasive and made it possible to think of using the same internet. Infrastructures like Facebook, in a different way to give everybody a Unique ID, an open-source kind of products, a software platform which built all kinds of applications on this unique ID.

Just like when Microsoft and Intel came together, they created a platform for PC – a transformation took place in the world of personal computers. Intel developed the chip, Microsoft developed the operating system, DELL or Compaq built the PC and brought a huge revolution of personal computers. The similar echo-system developed in the era of smartphones and tablets. On the other hand, ARM develops chips, which is licensed to a chip manufacturer media-tech in China or Qualcomm in the United States; Google or open-source software vendor provides the Android operating system and helped the whole system blossomed.

Example:

Mastercard, Deltek.

If we look back – information was scarce, so it was simple to handle to use as competitive advantage, but then there was transformation where things had become more complexed with lots of information around in different shapes and sizes and is increasingly growing more and more unmanageable – leading to "Death by Data".

So rather these data can be treated as strategic asset. Transforming all these information makes it an advantage.

Example: see Figure 3.3, a production house producing goods from raw materials and suddenly when the system breaks down, it may take a lot of time to replace the parts or fix the problem. So here we go for Information Transformation. Acting proactively used the information where the data can be collected, analysed and

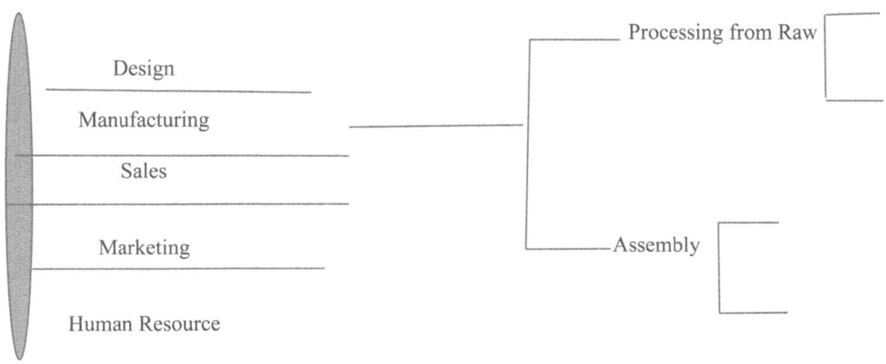

FIGURE 3.3 Organizational structure.

used to keep the businesses one step ahead to the mutual benefit of both production and finances, services marketing and sales.

Information transformation may help one to give the customer a systematic attention according to the needs and feedbacks and increase customers loyalty, customer satisfaction and boost sales. Information transformation helped one to create growth, manage risks and develop new innovative products.

3.3.4 Operating Model Transformation

A company turns input into output and makes money in the process; a manufacturer turns raw materials into finished or useful goods, services a retailer markets wholesale goods to the general public etc. However, an operating model(Ref Figure 3.4) is less about what a company does and more about how it does. It is very important to understand how the company is operating and getting where the value is being added to enhance the performance improvement activity to remove the waste which results in better efficiencies and profits. The organization model of a company.

These activities make the core process of the functions to add value to the business. It can have a detailed process map for the activity assembly; by showing lanes or swim lanes, two roles are involved in this process as drilling down or levelling. The level below is called as the sub-process and each activity can be expanded till the lowest level of details that needs to be represented. A process map is another dimension of an operating model. It describes how the company is performing its business activity by the actual processes that are undertaken by the people in the company, and an operating model can have a number of dimensions provided on how the company turns inputs into outputs. Organizational models and process maps are common examples. The real try is how the dimensions are linked with the operating model together, so that any change in one dimension essentially ripples through all other dimensions automatically without having to do any manual work even something as simple as a wrong name change. An operating model is about how a company turns its input into output. It can be constructed by any number of dimensions. It is important to understand where the value is being added and where performance improvement activity can be carried out in any of these dimensions.

An operating model is a document, a blueprint that helps managers to convert strategy into operational decisions. It helps managers on design and set up the

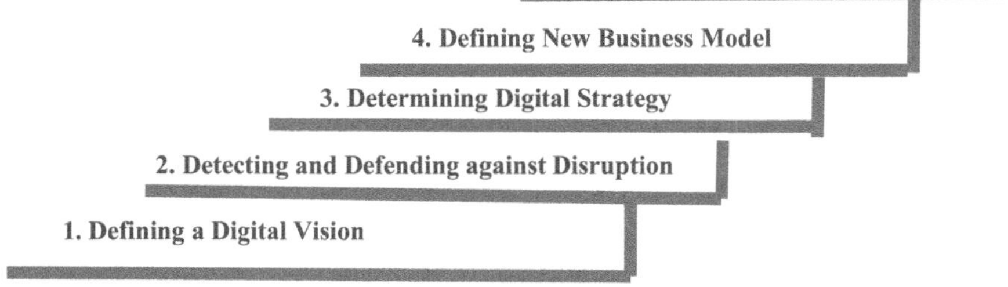

FIGURE 3.4 Five steps of digital transformation.

operations. It helps the head of operations – processes, supply chain processes and capacity needed. It helps the head of HR think about the organization structure, the targeted human assets that are required in the organization.

It helps the head of IT for IT architecture that will be needed to support the organization. So an operating model is a very important step between strategy and all the functional decisions that is to be taken in a particular area by the experts.

An operating model can be any number of documented pages, for example, the one of McDonald that can be thousands of documented pages to represent.

One page operating model is called an operating model canvas, also called business model canvas, which covers the back end of the business model canvas – activities, resources, partners that are part of the business model canvas. There are five alternatives that enhance the further development processes, suppliers, locations, organizations and informations. Working on operating model starts when there is a clarity about a value proposition that an organization is trying to deliver product or service to a customer, or internal functions similar to human resources that tries to deliver a compensation benefit plan for top management.

So it is very important to understand what value propositions to be laid out and delivered to suppliers, locations, organizations, people, information of software applications required to support these value propositions.

Example 1:

OLA – To provide low cost taxi services and also to provide rider, the client, who is using the taxi with more information about when that taxi is coming and what the price is going to be. So the software applications are the value proposition in the OLA platform that matches the riders to the drivers with a dynamic pricing model so when there are lots of drivers available. Such dynamic pricing model certainly helps clients get an appropriate price and ideally low price.

Example 2:

Cardboard citizens – A set-up for homeless people to rebuild their lives through improvised theatre. Through theatre they involve homeless people in producing theatre and give these people and insight into their life and abilities that can transform their lives. It has three value chains delivering different value propositions, all aimed at getting homeless people back into the mainstream of life.

3.3.5 Work-Source Transformations

Change is inevitable. All failure is a failure to adapt adaptability, and action creates opportunities. Work is no longer confined to a certain physical space. Each organization is a mini city itself with solutions at each touch points and maximizes the creativity, imagination and product activities of a workforce constantly pushing the limits of possibilities. Workforce built with diverse culture, ways of working and experiences all come together to achieve common goals. In a world, everything is connected and systems are reaching a state of perfection. The only thing that separates excellence for just good enough is the human mind and brain – the central point of all transformations of mobilization, digitalization and automation – amplifying each other for a perfect storm of change and this is achieved by workforce the humans.

Technology doesn't innovate, rather men do. Technology doesn't disrupt any industry, people do. Technology is enabling and amplifying the potential of people, which is far greater than before. The enormous potential of people combined aspirations with technology – fundamentally all is a change for betterment. There is a huge gap between the world we operate in and the potential of people and technology to do great things, and we confront it every day in our working lives.

Work – source transformation – optimizes the productivity flexibility of the internal and external contributors to organizational value. It helps one to identify the efficient resources to achieve business objectives, drive business outcomes by creating a modular, agile structure, facilitate relationship and maximize the productivity of interactions and facilitate the attitudes and beliefs that affect and reflect a more continuously connected and entrepreneurial workforce society.

3.4 COMPONENTS OF DIGITAL TRANSFORMATION

There are six different key elements for any digital transformation. They are as follows:

1. **Business-led technology roadmap**
2. **Talent**
3. **Agile delivery**
4. **Technology and tooling**
5. **Data management**
6. **Adoption and operating model change**

- **Business-led technology roadmap**
  It gives a clear picture of where the value is going to be and how to get them through a roadmap that gets short-handed as a strategy, but when compared to a traditional definition of a strategy, it's a business led usually through partnership between the business organizations and business leads, which is aligned across the leadership team of a company and it has a specified level that really focuses on the following:
  - What is the economic goal?
  - What is the upside wishing to be achieved?
  - What is it going to take to get there which can able to result in prioritized roadmap?
- **Talent**
  To ensure that it has the right skills and capabilities to execute and innovate.
  While referring to digital talent, which is the plan for the business executives:
  - What does it take to be able to develop a set of business leaders?
  - Who are able to work effectively hand in hand with technology leadership?
  - Who are able to unleash the potential of technology in their areas of the business?

- **Agile delivery**
  This is to increase the metabolic rate of the business. "Agile" is one of the most over-used words in business. The reality is, there are many that have agility in IT, or in pockets of technology. It's much harder to have agility that's truly cross-functional and that's truly enterprise wide and results in continuous delivery.
- **Technology and tooling**
  This allows one to move fast with flexibility and resiliency. It is important to enable the business-led roadmap. We see that technology will allow one to move fast with flexibility and resiliency which means core-system replacement.
- **Data management**
  To leverage data intelligence and derive a competitive advantage, companies have done a smaller scale digital transformation, in which data is important but it doesn't rise to the strategic level of importance of some of the others. However, to be able to operate a transformation, a broad scale across enterprise data ends up being a key unlock and if data is unavailable, then data governance, data structure, data organization get tangled and can't set up in the right way. It can get into a tangled situation and will be unable to go from applications of digital-to-real drive value.
- **Adoption and operating model change**
  To evaluate value captured by ensuring mindsets and operational shifts, both internally and externally users unlock the value.

3.5 STEPS TO DIGITAL TRANSFORMATION

Digital transformation helps one to navigate the world because it is not just about putting in another system:

- How to do business in this digital world.

The physical and digital transformations are both combined together to do any business. In the 21st century economy, every business is actually operating in the digital world. It means that company must evolve. Company must just not think about digital transformation as just putting IT type of system. However, way of doing, thinking and understanding the way the customer have evolved to survive and thrive in a physical and digital world, which gives rise to a clear digital transformation outcome.

The five steps of digital transformations are depicted in Figure 3.4.

4 Artificial Intelligence for Smart City Development

Review/Survey of Existing Systems in Smart Cities

Vijayalakshmi A. Lepakshi[1] *and*
Prashanth C. Ranga[2]
[1]REVA University, Bengaluru, Karnataka, India
[2]Conestoga Institute of Technology, Kitchener, Ontario, Canada

4.1 INTRODUCTION

Technology and modernization are attracting more and more people to migrate to urban areas and cities, making those places highly populated, which leads to an imbalanced social ecosystem. By 2050, it is expected that 70 per cent of the world's population will live in cities, which will face challenges from sustainability to service delivery (Sun et al., 2016). Rapid urbanization increases challenges that need to be addressed, such as transportation, healthcare, education, energy and social cohesion. Technological advancements and information and communication technology (ICT) have emerged as solutions, turning a city into a smart city (SC). The term "smart" indicates skilfulness of various integrated things (Kuru and Ansell, 2020) to achieve the mission of a smart home (sHome), smart building (sBuilding) etc. Technologies such as wireless communication, Internet, Internet of Things (IoT), Artificial Intelligence and Block Chain have enabled smart applications to grow vast and diverse (Yang et al., 2021). Hence, SCs are expected to provide promising business solutions, leading to growth in the economy.

As demographic and environmental conditions of each city vary, the services provided to the stakeholders and the expectations and outcomes also vary. A properly built SC will have a different value for different stakeholders (https://strategyofthings.io/smart-city-ecosystem). Solution providers think of using technology for smart parking, smart lighting etc. City officials think of providing online services to citizens, such as applying for permits or searching online records. City residents expect smart mobility, a safe and secure environment, crime reduction, smart healthcare etc. As an SC is constructed on the basis of technology with a focus on outcome, cities worldwide are adopting the SC concept as a technological solution. The proposed SC projects are different in developed countries from developing countries and make use of the integrated technologies such as IoT, cloud computing, edge computing, bigdata and analytics and so on (Beretta, 2018; Calzada, 2018; Camboim et al., 2019;

 DOI: 10.1201/9781003225317-4

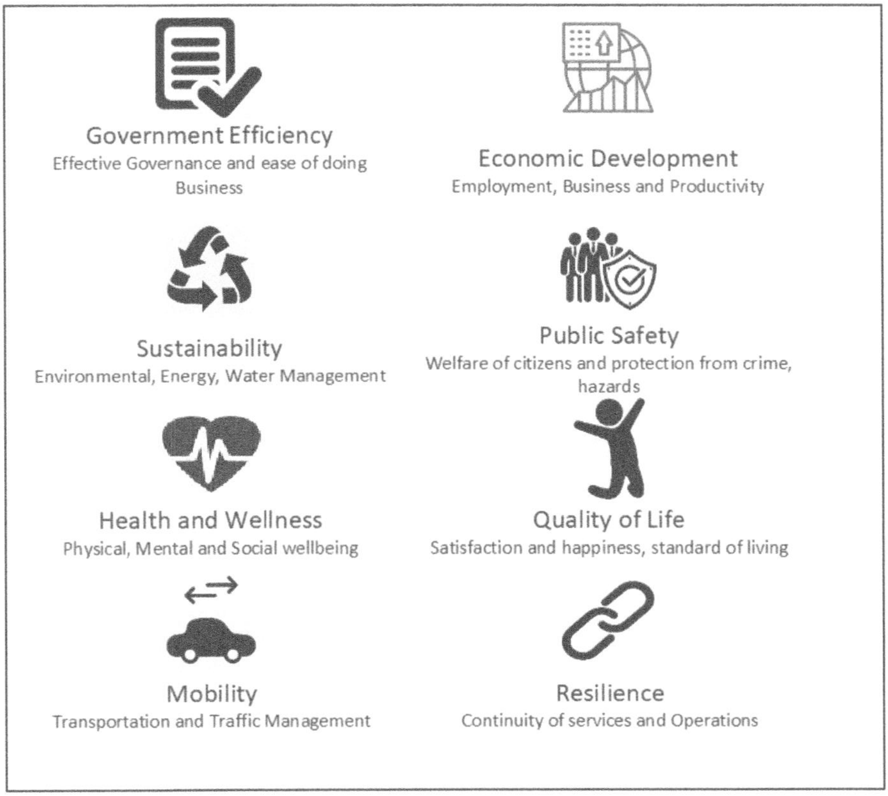

FIGURE 4.1 Outcomes of smart city projects.

Chiodi, 2016; Desdemoustier et al., 2019; Lytras, 2018; Sikora-Fernandez, 2018). The probable outcomes of SC projects are depicted in Figure 4.1.

As the existing SCs have evolved as a solution to urban problems, the definitions of an SC vary and rely on parameters such as key values of SCs and services offered. The Institute of Electrical and Electronics Engineers (IEEE) defines (https://strategyofthings.io/smart-city-ecosystem) "a smart city as one that brings together technology, government and society to enable the following characteristics: a smart economy, smart mobility, a smart environment, smart people, smart living, smart governance". Cisco envisions an SC as "a city with a pervasive overlay of information and communication technology (ICT) connecting things, organizations, and people". Cisco calls this future vision Internet of Everything (IoE) (Cisco).

Mohammad et al. (2019) stated the definition given by the International Telecommunication Union (ITU) about an SC: "A smart sustainable city is an innovative city that uses information and communication technologies (ICTs) and other means to improve quality of life, efficiency of urban operation and services, and competitiveness, while ensuring that it meets the needs of present and future generations with respect to economic, social and environmental aspects". Hence, the SC's vision is to improve liveability, conservation, renaissance and sustainability with an aim of making a community to live in present, remember the past and plan for future of a community (Sun et al., 2016).

The rest of this chapter discusses the smart city ecosystem, various services offered by the smart city, and technologies used for building smart cities. In addition, it also discusses a few case studies showing the successful implementation of smart city as well as the challenges to be faced in building smart cities.

4.2 SMART CITY ECOSYSTEM FRAMEWORK

A sustainable SC ecosystem integrates people, organizations, policies, laws and processes together to produce the desired outcomes. This city should be adaptive, receptive and appropriate to all the citizens who live and work there, as well as those who visit the city. In an SC scenario, technology plays a major role and is integrated to quickly enable and transform this ecosystem. An SC is built on multiple layers that provide various capabilities, where each capability contributes to achieve the required outcome (https://strategyofthings.io/smart-city-ecosystem). No capability can be considered more important than others, when technology is one of the required capabilities for deploying SCs. The various components involved in the SC ecosystem framework are depicted in Figure 4.2.

- SC technology infrastructure layer – A scalable SC technology infrastructure is required in providing services to the stakeholders.
- Connectivity, accessibility and security layer – The ability to connect people, things and systems in a seamless fashion through a trusted connection.

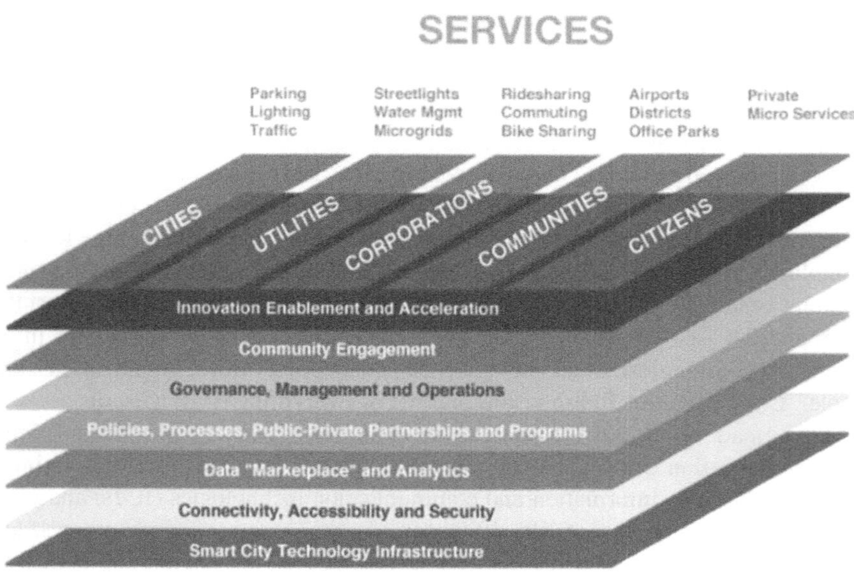

FIGURE 4.2 A framework for the smart city ecosystem.

- Data marketplace and analytics layer – SCs depend on data market places, open data, and analytics. It is equally important to have programmes that share and process data in a secured way.
- Policies, processes, public-private partnerships and programmes layer – The businesses, financing sources, rules and policies and so on are organized in this layer.
- Governance, management and operations layer – Government organizations must upgrade themselves to propose, provide and invest in new business models.
- Community engagement layer – Community involvement in planning and implementing the services and for assessing and improving existing services is a critical issue.
- Innovation enablement and acceleration layer – Service providers should involve continuous innovation and upgrading services for stakeholders.
- Value layer – This layer is visible for the stakeholders, which provides various services transparently.

The process of developing an SC is a continual process and an SC ecosystem advances using well-established capabilities such as plug-and-play, which support deep-rooted integration plans and policies. SCs are developed with the objective to build unsupervised automatic systems with little or no human intervention required while increasing the quality of life.

4.3 SMART CITY SERVICES

In an SC environment, stakeholders' requirements vary based on the demographic, social, economic and environmental conditions. Yang et al. (2021) examined various services offered by the SCs across the world. Based on the common services offered across these cities, six service domains are characterized: Living, governance, mobility, built infrastructure, industry and natural resources. Broad categorization of service domains and sub-domains is represented in Figure 4.3.

Based on these broad domains, the varied services provided for the citizens are smart governments, smart healthcare, smart transport and mobility, sHome, smart energy, smart community, smart shopping, smart education, smart lighting, smart waste management and smart business, to name a few, but they are not limited to these services. A few common services that are offered in many SCs worldwide are discussed in the following sections.

4.3.1 Smart Governments

In an SC environment, users' needs vary, and efficient services designed to meet the needs of the public act as the driving force. The SC should take the responsibility to provide services and support for the public round the clock. The smart government should (a) effectively integrate all service domains, (b) coordinate among all government departments in the city along with the central government through these integrated service domains, (c) provide access to all IT services through official

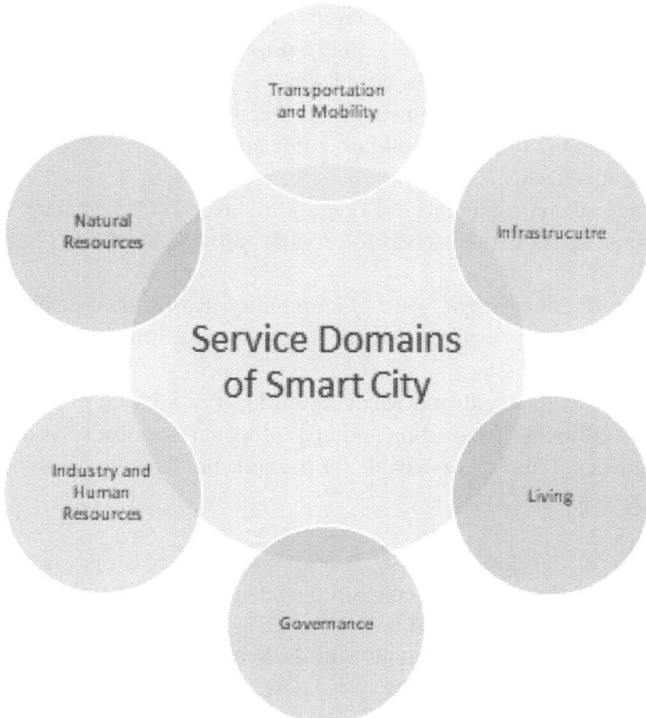

FIGURE 4.3 Broadly categorized service domains and sub-domains.

government websites, for example, applying for school admissions and elderly care services at home, paying taxes etc and (d) provide information and early warnings using SC applications during emergency situations (Kuru and Ansell, 2020). Smart government should aim at offering public services at reduced cost with enhanced mechanisms to manage a city in a more transparent, unbiased and affordable way by using real-time data and feedback.

4.3.2 Smart Mobility

Multimodal smart mobility is provided to users "as a service" that solves their mobility needs successfully and efficiently (Al-Rahamneh et al., 2021). This system uses data produced by IoT systems and data generated and received from other sector applications in the city. This information contains the schedule for public transportation such as buses and trains, the availability of parking slots in public places among various parking places in the town, the warning messages sent from the civic government and the information received from local traffic control rooms about traffic jams and many more. Subsequently, data processing is performed and interoperable context information is produced as per the relevant context for smart applications. Finally, the extracted information is used for development of the decision-making process, and simultaneously information is provided to the user.

4.3.3 Smart Healthcare

SCs utilize ICT to provide various services to citizens in the healthcare sector using mobile technology and IoT sensors. These sensors collect ambience information and combine with machine learning and Big Data technologies for further processing and decision making (Cook and Duncan, 2018). SCs can gather information from various sources, in addition to data fetched from citywide locations such as the power grid, transportation, vehicular networks, locations of emergency service providers and magnitude of crowds in locations throughout the region. Smart devices such as phones and watches are equipped with various sensors to collect data; for example, common data to be collected is location information using a combination of mobile computing technologies such as GSM, GPS and Wi-Fi, depending on device's location, whether it is inside or outside a building. Additional information can be collected from various components such as cameras and microphones to provide sources of data that indicates the state of the user and environment based on the actual device that is used. Information can also be fetched using installed applications on the device, including phone call apps and texting apps. For example, detecting medical emergencies using sensors located in a smart home, such as a person with Parkinson's falling in the kitchen may be a probable medical emergency detectable using ambience information.

4.3.4 Smart Energy

A smart grid (SG) comprises various components such as a control device, electrical and electronic devices like phasor measurement unit (PMU), smart meter, information acquisition system and distribution unit etc. Smart energy or SG networks are key components in building SC architecture towards constituting a sustainable energy for future. They facilitate the integration of renewable energy sources and electrification of transport. They also enable new energy-related, value-added services (Masera et al., 2018). The key technical elements to be considered in the implementation of SGs are as follows: (a) integration of distributed renewable energy sources, (b) grid controllability and active management, (c) management of demand at different time scales and (d) convergence to common standards. In terms of sustainable development, focus should be on potential climate and environmental impact considering air quality improvement, reduction of CO_2 and other greenhouse emissions etc.

4.3.5 Smart Community

A group of people who live within a neighbourhood is called a community. Communities with localized needs can be considered as mini-SCs. Potential smart community examples include huge campuses such as at a university, an office park, an airport, an apartment complex, a housing development/neighbourhood, a business district and even individual "smart" buildings. These communities need specific tailor-made smart services for their citizens. Even though more than 50 per cent of the population lives in the urban areas, the remaining people live in rural areas and hence the smart community should take urban as well as rural populations into account for a community service to be beneficial for more people (Sun et al., 2016). Smart communities should envision improving liveability in terms of providing adequate transportation choices, promoting affordable housing, enhancing economic competitiveness, supporting existing communities by coordinating with policies, leveraging investment and valuing communities and neighbourhoods. Liveability should be one of the prime characteristics for smart communities that thrive to preserve cultural and natural heritage. Thus, the definition of sustainable development can be stated as "the development that meets the needs of the present without compromising the ability of future generations to satisfy their own needs". The community is said to be sustainable if it maintains its economic and environmental health, promotes social equity and fosters broad citizen involvement in planning and implementation.

4.3.6 Smart Education

Technology transformed the education sector in a broader way. Smart boards, laptops, tablet devices, e-readers, mobile devices and even e-learning apps have become part of everyday education while replacing the traditional classroom environment with a chalk and blackboard. In contrast to the traditional teaching model, flipped classroom model came into existence, in which teachers provide pre-recorded lectures outside of classroom through online platform in the form of MOOCs, and classroom time is effectively utilized for implementation and practice. In this system, prior to attending classes, students are allowed to learn at home at their own pace, supported by the teacher through video recordings or online chats and then class time is used to discuss the lesson and learn through activities and interaction (Mohammad et al., 2019; Zhu et al., 2016). A technology-supported environment can generate smart students.

4.3.7 Smart Lighting

The optimization of streetlight efficiency is one of the significant features in SC deployment. This service optimizes the intensity of the streetlight, particularly depending on the time of the day, the weather condition and the presence of people (Zanella et al., 2014). Furthermore, through a mobile application, provisions can be made to operate streetlights as per the requirement of the users; hence, to provide such service the streetlights need to be included in the SC infrastructure. Moreover, a fault detection system can also be easily implemented along with the streetlight controllers.

4.3.8 Smart Waste Management

In many modern cities, waste management has become a primary issue, due to the cost incurred to provide the service and the problem of storing garbage in landfills. In this domain, an ICT solution may result in substantial advantage economically and ecologically. For instance, by using intelligent waste containers, the load level of the truck can be detected and the collector truck's route can be optimized, which leads to the reduction of waste collection costs and improved waste recycling. Thus, smart waste management service can be implemented by connecting the intelligent waste containers to a control centre as end devices with the help of IoT sensors, in which data processing takes place by the optimization software and governs the optimal management of the collection truck fleet.

4.3.9 Smart Home (sHome)

sHome or home automation can be deployed using a few industry standards that provide an easy, economical, happier and efficient day-to-day life with the help of connected IoT devices using wireless technologies (Sun et al., 2016). An automated sequence of events that forms a pattern – such as closing curtains and switching off television, music system and/or room heating – can be triggered while leaving home to make the life easier. Other examples are opening curtains heating up the house, etc., while on the way back home; switching off the television and lights before going to sleep; switching on a preferred TV channel and the coffee maker upon waking up; all this can be triggered in a smart home. In sHome deployment, energy efficiency is one of the major concerns. By using location-independent remote monitoring and control abilities, the automated devices at home can be managed from any place. For example, while being at work place, a washing machine at home can be turned on, lights at home can be turned on/off or one can open/close curtains while enjoying a holiday away from home and so on. As sHome is located in an SC, it should be integrated with all the SC domains such as: (a) smart security, for example, monitoring the environment in and around a house, which may trigger a warning for the police or houseowner; (b) smart community, for example, informing and training citizens at their homes; (c) smart energy, for example, smart metering, smart control switches etc; (d) smart health, for example, monitoring the health of elderly people or patients at home remotely, which may trigger an ambulance service and/or doctor actuation; and (e) smart environment monitor, for example, sending alerts under emergency situations such as floods or earthquakes. In sHome, alarms installed for fire or gas may trigger an alert for police, fire department or ambulance service in an SC in emergency situations, ensuring safety for the public.

4.4 TECHNOLOGIES FOR SMART CITY DEVELOPMENT

Technological advancements in the recent era have paved the way for improving living environments of citizens. In SC implementation, various entities are interconnected, including technology. In an SC framework, the underlying technology and its role can be analysed as depicted in the following sections.

4.4.1 Information and Communication Technology (ICT)

ICT plays a crucial role in the SC framework by making efficient use of physical infrastructure, resources for adapting changing circumstances and connecting citizens with various services (Andrisano et al., 2018). An SC can be realized by interconnecting various entities such as data generation, data processing and end-users. The main components in ICT infrastructure are the communication component, processing component and storage component, which help to realize the requirements of an SC by viewing them as a framework to represent a technological nervous system of the SC to afford a better urban life for the citizens. The interconnection of IoT infrastructure and its core components of SC is depicted in Figure 4.4.

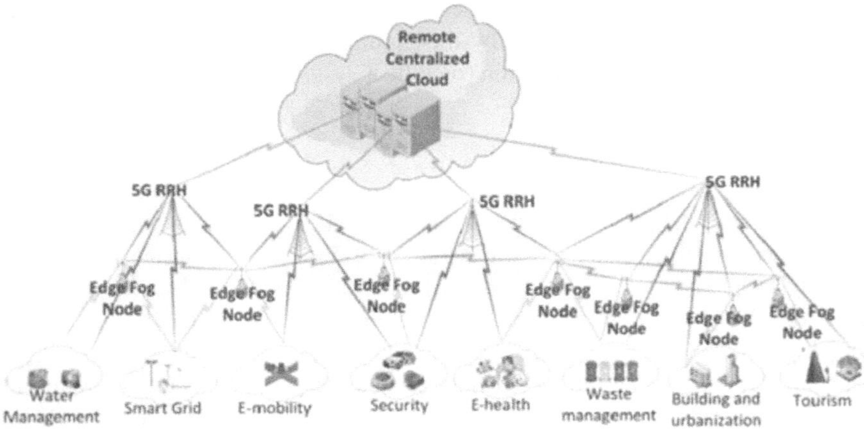

FIGURE 4.4 Interconnectivity between ICT infrastructure and smart city core components.

4.4.2 Internet of Things (IoT)

SCs gather a large amount of data from various sources. The sensors are able to gather various types of data regarding users and their environment. In an IoT communication framework, the objects of everyday life are equipped with microcontrollers, transceivers for digital communication to communicate with other connected devices and users through a protocol stack which is an integral part of the framework. IoT fosters a development of various applications that makes use of enormous data generated by various objects to provide services to citizens, organizations and public administration in many different domains such as home automation, industrial automation, healthcare, elderly assistance, SGs, automotive, traffic management and many others. The urban IoT infrastructure has the capability to integrate different technologies with the existing communication technologies and also make the collected data available for users and authorities, making them responsive towards cities' problems. The various devices that are part of the urban IoT infrastructure are as follows: (a) IoT nodes such as sensor/actuator equipment, nodes with powering mode or networking role and RFID tags; (b) mobile devices and smart devices that provide other ways to interact with connected equipment; (c) gateway devices that play a major role for establishing interconnectivity among end devices providing protocol translation between constrained and unconstrained counterparts such as XML-EXI and HTTP-CoAP; and (d) back-end systems for interfacing with IoT systems such as database systems, websites and ERP systems.

4.4.3 5G Technology

5G technology is a fifth-generation mobile network system that operates at data transmission speeds over 20 GBPs. It supports significant broadband speeds and heavier data usage than previous generations while enabling the full potential of IoT. A 5G network has the ability to connect a million devices per square kilometre and can cater to real-time communication for applications such as autonomous vehicles and remote patient monitoring. Security standards in 5G stress system confidentiality, integrity and encryption in both user equipment and 5G network infrastructure. IMSI encryption implemented within the 5G sim delivers improved security, control and interoperability among connected devices.

4.4.4 Cloud Computing

Cloud Computing is a distributive environment, and a cloud platform can be expandable vertically in terms of massive storage, huge computation power and global connectivity. In an SC scenario, IoT devices are used for data collection of users and their environment. In general, IoT devices come with limited processing and storage capabilities, whereas SC analyses a huge amount of data collected from various sources. The cloud infrastructure can be used as a solution for storing and processing the collected huge amount of data from these connected devices. Cloud can deploy virtual machines (VMs) in the wide networks with smaller versions of cloud platforms known as cloudlets using hypervisors that enable real-time processing (Kuru and Ansell, 2020).

4.4.5 Big Data Analytics

In an SC, a vast amount of data, called *Big Data*, is collected from millions of sensors and other interconnected devices, demanding further processing to provide intelligence to stakeholders. This Big Data that represents the real world increases the chance of extracting better insights for decision-making (Kuru and Ansell, 2020). Big Data Analytics enable the real-time applications for various application domains such as healthcare, buildings, traffic, automotive and crime prevention to name a few (Zanella et al., 2014). Apart from all the previous examples, data gathered from SC components will be heterogeneous, and translating this biological, physical and social data into a meaningful form is further a challenging task. For example, raw data collected from sBuildings includes room occupancy data, outdoor data such as weather, data related to energy consumption, data related to state of the doors and windows, airflows, building structure and materials and so on. As part of the Big Data Analytics process, erroneous and noisy data is cleaned to transform to useful data and unnecessary data is filtered out. To handle heterogeneity of data, various issues that need to be addressed are improving the accuracy of real-time data, data representation and processing models, improving interpretation for interoperability, knowledge mining and reasoning and short- and long-term storage of data. Hence, Big Data Analytics can probably be analysed in various levels such as finding (a) **insight** – deep understanding of trends in the data and statistics, (b) **outsight** – understanding of societal and external aspects, for example, correlating events and situations with weather influence and (c) **foresight** – predictions for decision-making.

4.4.6 Edge Computing

Edge computing is an emerging architecture used in SC deployment that provides computing, storage and networking services closer to end-users (Kuru and Ansell, 2020). Edge computing provides high privacy and security and enables independent processing and organization of real-time applications that require decision-making with low latency, through mobile services. This edge computing technology is particularly beneficial for real-time operations, rapid innovation and services such as fire detection and video surveillance.

4.4.7 Artificial Intelligence

Artificial Intelligence has gained popularity in providing intelligence services in all fields. Machine learning and deep learning techniques are used extensively in intelligent decision-making to provide services in various SC domains. These techniques make the applications smarter in providing recommendations in smart shopping, smart healthcare, SG, sHome, smart mobility and so on. Hence Artificial Intelligence is a ubiquitous technology which is extensively used in SC deployments.

4.5 CASE STUDY ON SMART CITY SERVICES

Technological advancements motivated visionary organizations and governments to plan and implement smart services for citizens based on their requirements to improve the quality of life. A few successfully deployed SCs and the services provided in these are discussed in the following sections.

4.5.1 Lake Nona

The first Smart plus Connected city designated by Cisco is *Lake Nona*, which is located in Florida, USA, demonstrating the enabling of next generation citizen services by the network (Cisco). Lake Nona is a 7000-acre planned community that has been developed by an international investment firm, the Tavistock Group, within the city limits of Orlando. Lake Nona is a smart community with residential homes and the Lake Nona Medical City – a life science cluster of medical schools, healthcare providers and research centres.

Lake Nona project is a US $2 billion project taken up by the Tavistock Group with a vision to brand Lake Nona as a global role model for other cities, in which the technology infrastructure is viewed as an essential utility similar to water or electricity. Cisco being part of the project enabled services such as unlimited connectivity, computers, storage and bandwidth.

The leadership model implemented in Lake Nona is based on a collaboration model that will foster innovation to realize the vision. Leaders from hospitals, research centres and various other councils are brought together to identify the future needs of Lake Nona and CIOs, and representatives of various councils provide input on how to use technology to meet needs. Various organizations, instead of operating independently, brought expertise and interests together to prioritize the highly valued projects for the citizens and focused on rapid deployment.

Cisco became a partner of these councils to coordinate on how to use the technology infrastructure for pilot solution areas focused on healthcare, education, community services, safety and security, intelligent workspaces, innovation and experience while working together in their expertise areas.

For example, the impact of technology on the lives of Lake Nona residents is demonstrated through two pilot projects as given below:

Healthcare: In Lake Nona a patient-centric healthcare system is implemented. Lake Nona enables devices at home to connect directly to hospital monitoring systems. A doctor remotely visits the patient at home using "Cisco's HealthPresence solution". For example, a remote doctor monitors the patient's heartbeat while interacting via a video call.

Learning and education: Lake Nona High School pioneered implementing the concept of a flipped classroom. The flipped classroom is different from the traditional teaching model in terms of delivering content. In the flipped classroom model, the instructions or content are delivered in advance outside of the classroom, via pre-recorded lessons online and the class time is effectively utilized for the discussions and practising. In this method, students study at home at their own pace, supported by video recordings or material and online communication through chats to talk with peers and the teacher, and then class room time is used for lesson discussion and learning through activities and interactions.

4.5.2 Padova Smart City

Padova SC is a practical implementation of an urban IoT with a private-public collaboration such as the Padova municipality, the project sponsor, with the Department of Information Engineering of the University of Padova, which provided the feasibility analysis of the project and Patavina Technologies that developed the IoT nodes and the control software. Zanella et al. (2014) proposed and implemented urban IoT with a goal of promoting early adoption of open data and ICT solutions for public administration. Padova SC architecture has various hardware and software components and collects open environmental data for decision-making as depicted in Figure 4.5.

The hardware and software components include streetlights, constrained link layer technologies, WSN gateways, HTTP CoAP proxies, database servers and operator mobile devices. Streetlights are geographically connected to IoT nodes and are equipped with sensors for collecting environmental data such as temperature, humidity and light intensity. IEEE 802.15.4 link layer technology is used for providing routing functionalities via the IPv6 Routing Protocol for Low power and Lossy

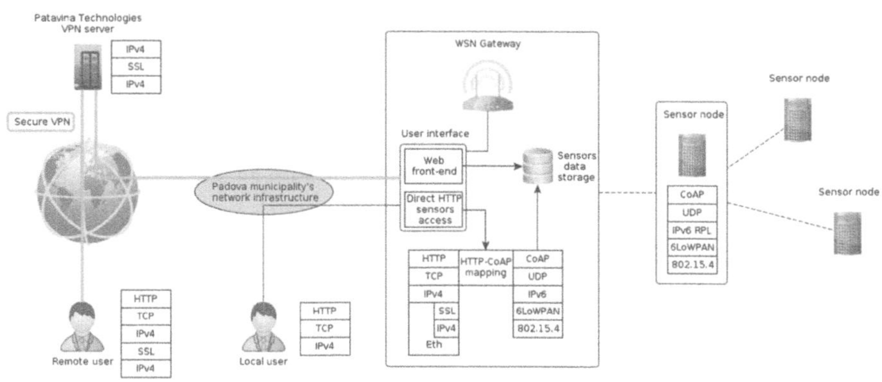

FIGURE 4.5 System architecture of the Padova smart city.

Networks (RPL). WSN gateways work as an interface to cloud sensors and central back-end servers. HTTP CoAP proxy enables communication with CoAP transparently either by polling technique or by subscribing technique from the node only when the measured value by the sensor falls out of range. The database server collects information from connected nodes, which is accessible from web programming technologies. In a Padova system, the database server is realized on the WSN Gateway, allowing plug and play connectivity. Public lighting operators are equipped with a mobile device to locate the streetlights with issues. Hence, urban IoT is implemented to realize the SC concept in a Padova smart city.

4.5.3 The South Korean National Strategic Smart City Program

The National Strategic Smart City Program (NSSP) was launched in South Korea with unique characteristics with an aim to make cities sustainable and widen industrial territories with public and private associations (Yang et al., 2021). The NSSP covers all six major domains: living, transportation, natural resources, infrastructure, governance, economy and education. The NSSP focused on three major projects to realize the goals of an SC. The focus of the first core project is to develop source technologies, that is, infrastructure that includes "data architecture" and "networking protocol," which thus does not provide any particular service. The focus of the second core project, which involves the Daegu metropolitan area, is more on infrastructure management and mobility service. The focus of third core project is the electricity producer KEPCO, involving the relatively small city of Siheung. In these projects, major concerns were energy management, air quality and living environment management.

The focus of NSSP in the transportation and mobility domain is to demonstrate the significance of usage of autonomous vehicles. One of the core projects of NSSP is implemented in Siheung City, which provides a pioneering level of mobility for citizens that includes autonomous vehicles for smart mobility technologies as a sandbox. Siheung's urban area is spread into small towns, where one's own mobility is preferred for moving around the city. Another core project of NSSP is implemented in the city of Daegu to develop a prediction system to provide mobility as a service, sensitive to the demand for spatial transportation. Another main service in this domain is the development of a smart parking service and parking data structure. Parking is a persistent problem in old and dense cities, and providing parking information to public is a common service in SCs globally. As a solution, the smart parking data structure is standardized to create another SC business, and a parking management system was developed. Even though the projects are implemented in Daegu and Siheung, the project is intended for all the Korean cities.

4.5.4 TreSight – Trento, Italy

TreSight (Sun et al., 2016) is context-aware recommender system for the development of sustainable cultural heritage and smart tourism in the city of Trento, Italy. To provide smart tourism service, a wearable bracelet that can communicate with a mobile app is used to collect contextual real-time data to be used with static data coming from Open Data Trentino about the point of interest, restaurants etc. Wearable bracelets are used to crowd-sense the activity of citizens or tourists, environmental conditions such as temperature and humidity and success rate in terms of number of tourists visited by tracking the wearable bracelet. Places of interest gather and make available information about availability in tourist places, tourist density and functioning hours of points of interest or places by detecting the on or off status of the hotspot.

This kind of solution can be built with a collaboration model. The TreSight project is partially funded by the city tourism department and the sponsorships received by theme parks, restaurants, pubs, museums and other interested places to advertise or promote themselves. The end-users will benefit from the services provided by using the open data and getting context-aware recommendations. Further, the mobile app can be enabled for discovering friends in the city. Hence, TreSight provides context-aware recommendations for smart tourism and sustainable cultural heritage further leads to growth in the economy.

4.5.5 STARDUST Project

The STARDUST project is aimed at building highly efficient, smart and citizen-oriented cities with low carbon. The STARDUST project initiates developing urban solutions and new business models that integrate the domains of buildings, mobility and green energy through ICT.

4.6 CHALLENGES IN BUILDING SMART CITIES (SCs)

Citizens play an important role in building SCs. Building an SC is a constant process and ever-changing demands of the growing urban population pose new challenges time to time. Also, the new advancements in technology will pose various implementation and integration problems with all entities in the SC ecosystem. Engaging people with these services is another challenge due to lack of digital skills of residents and lack of smart devices due to economic conditions. Apart from these, the various challenges to be faced by the people who adapt these services are discussed in the following sections.

4.6.1 Cybersecurity

SCs can be human-centric systems; hence, preserving the security and privacy of an individual is important (Sun et al., 2016). Crowdsensing techniques used in SC applications gather sensitive information relating to individuals through sensors. For example, traffic congestion levels can be estimated by GPS sensor readings in a given community. Meanwhile this information can also gather private information about the individuals, such as home and work locations, as well as the daily commute routes. Security and privacy policies influence the management of accessing source information, recording ownership of data and processing history of data within public services.

4.6.2 Privacy Protection

In an SC, privacy-sensitive information is accumulated by applications from citizens and their social contacts that raise concerns over security and privacy. Applications of either autonomous systems or involving human intervention control city services and influence lives of citizens (Sun et al., 2016). For example, an intelligent surveillance system actually intended for monitoring unlawful behaviours in real as well as cyber worlds may gather local residents' day-to-day life patterns, lifestyles or even privacy. In this scenario, the data should be anonymized before publishing for public use to ensure privacy and security. A sanitization technique known as anonymization/de-identification is used to ensure non-traceability to an individual or household. A new technology known as "Block Chain" has recently been introduced to reduce these apprehensions in SCs.

4.6.3 Resource Limitations

The mobile devices are highly dynamic in terms of availability, and capabilities are used for collecting sensor data from various sources. The prediction of energy and bandwidth requirements for applications used for an SC is much harder due to its highly dynamic nature than traditional wireless sensor networks. Apart from this, tasks that need categorizing, scheduling, sensing and communication cannot be completed due to varying demands and usage patterns by the stakeholders.

4.6.4 Heterogeneous Data

A huge amount of data gathered from SC components will be heterogeneous and hence cannot be used as it is, and translating this biological, physical and social data into a meaningful form is further a challenging task. To handle the heterogeneity of data, various issues need to be addressed for improving the accuracy of data in real time, data representation and processing models, improving interpretation for interoperability, knowledge mining and reasoning and short-term and long-term storage of data.

4.7 SUMMARY

Globalization and urbanization are leading to the growth of population in the cities leading to imbalance in the environmental, economic and ecological ecosystem. The ever-increasing demands of the cities that are equipped with smart infrastructure to provide varied services continue to emerge with the increasing urban population posing new challenges. Worldwide several SCs have been established that offer services for various smart domains based on demographic and environmental conditions. The services provided by an SC include smart governance, sHome, smart mobility, smart education, smart healthcare to name a few. Hence there is no unique SC framework that can be considered an ideal framework that provides all services for establishing SC in the future. To convert cities to SCs, centre and local governments should initiate the funds but with a public involvement and support in a collaborative model. A sustainable SC ecosystem involves organizations and businesses, policies, citizens, laws and processes to integrate together in creating the desired outcomes that satisfy the requirements of all stakeholders. By harnessing smart technologies, an SC can be developed as a more sustainable city that mitigates the rapid urbanization challenges. Hence the goal of building SC should be improving liveability, which enhances quality of life of citizens and a sustainable future for future generations.

REFERENCES

A. Al-Rahamneh et al. (2021). "Enabling Customizable Services for Multimodal Smart Mobility with City-Platforms". *IEEE Access*, vol. 9, pp. 41628–41646, doi: 10.1109/ACCESS.2021.3065412.

O. Andrisano et al. (April 2018). "The Need of Multidisciplinary Approaches and Engineering Tools for the Development and Implementation of the Smart City Paradigm". Proceedings of the IEEE, vol. 106, no. 4, pp. 738–760, doi: 10.1109/JPROC.2018.281.

I. Beretta (Feb. 2018). "The Social Effects of ECO-Innovations in Italian Smart Cities". *Cities*, vol. 72, pp. 115–121.

I. Calzada (Sep. 2018). "(Smart) Citizens form Data Providers to Decision-Makers? The Case Study of Barcelona". *Sustainability*, vol. 10, no. 9, pp. 3252–3276.

S. I. Chiodi (Jul. 2016). "Crime Prevention through Urban Design and Planning in the Smart City Era: The Challenge of Disseminating CP-UDP in Italy: Learning from Europe". *Journal of Place Management and Development*, vol. 9, no. 2, pp. 137–152.

Cisco (2013). https://www.cisco.com/c/dam/en_us/solutions/industries/docs/scc/ioe_citizen_svcs_white_paper_idc_2013.pdf. (Accessed in January 2022).

D. J. Cook and G. Duncan (April 2018, Epub 2018 Jan 23). "Using Smart City Technology to Make Healthcare Smarter." Processing of the IEEE Institute of Electrical and Electronics Engineers, vol. 106, no. 4, pp. 708–722. doi: 10.1109/JPROC.2017.2787688.

G. F. Camboim, P. A. Zawislak, and N. A. Pufal (May 2019). "Driving Elements to Make Cities Smarter: Evidences from European Projects". *Technological Forecasting and Social Change*, vol. 142, pp. 154–167.

J. Desdemoustier, N. Crutzen, and R. Giffinger (May 2019). "Municipalities' Understanding of the Smart City Concept: An Exploratory Analysis in Belgium". *Technological Forecasting and Social Change*, vol. 142, pp. 129–141.

J. Yang, Y. Kwon, and D. Kim (2021). "Regional Smart City Development Focus: The South Korean National Strategic Smart City Program". *IEEE Access*, vol. 9, pp. 7193–7210, doi: 10.1109/ACCESS.2020.3047139.

K. Kuru and D. Ansell (2020). "TCitySmartF: Comprehensive Systematic Framework for Transforming Cities into Smart Cities". *IEEE Access*, vol. 8, doi: 10.1109/ACCESS.2020.2967777.

A. V. Lytras (Aug. 2018). "It's Not a Fad: Smart Cities and Smart Villages Research in European and Global Contexts". *Sustainability*, vol. 10, no. 8, pp. 2727–2736.

M. Masera, E. F. Bompard, F. Profumo, and N. Hadjsaid (April 2018). "Smart (Electricity) Grids for Smart Cities: Assessing Roles and Societal Impacts". Proceedings of the IEEE, vol. 106, no. 4, pp. 613–625.

N. Mohammad, S. Muhammad, A. Bashar, and M. A. Khan (2019). "Formal Analysis of Human-Assisted Smart City Emergency Services". *IEEE Access*, vol. 7, pp. 60376–60388, doi: 10.1109/ACCESS.2019.2913784.

D. Sikora-Fernandez (Aug. 2018). "Smarter Cities in Post-Socialist Country: Example of Poland". *Cities*, vol. 78, pp. 52–59.

Y. Sun, H. Song, A, J. Jara, and R. Bie (2016). "Internet of Things and Big Data Analytics for Smart and Connected Communities". *IEEE Access*, vol. 4, pp. 766–773, doi: 10.1109/ACCESS.2016.2529723.

A. Zanella, N. Bui, A. Castellani, L. Vangelista, and M. Zorzi (2014). "Internet of Things for Smart Cities". IEEE *Internet of Things Journal*, vol. 1, pp. 22–32, doi: 10.1109/JIOT.2014.2306328.

Z.-T. Zhu, M.-H. Yu, and P. Riezebos (2016). "A Research Framework of Smart Education". *Smart Learning Environments*, https://doi.org/10.1186/s40561-016-0026-2.

5 AI-Enabled Smart City Waste Management System

Prakash Kumar Udupi[1], *Manju Jose*[2], *and Asad Ullah*[1]
[1]Middle East College, Muscat, Oman
[2]Majan University College, Muscat, Oman

5.1 INTRODUCTION

A smart city facilitates the broad integration of technology to enhance the quality of life of citizens in urban environments. Managing solid waste, which consists of discarded consumable waste, household waste or food waste in urban environments, needs diversified infrastructure, proper planning and integration of relevant technologies for disposing the waste. Waste disposals and waste recycling are among the essential needs of the smart cities, which also need to comply with regulatory requirements. Technology-enabled waste management systems such as artificial intelligence (AI), IoT (Internet of Things), big data, machine learning and smart sensors provide long-term and sustainable smart city waste management frameworks. Modern urban smart city waste management systems also need to address next-generation technology breakthroughs, such as waste to energy, automated waste sorting systems, effective usage of IoT to increase efficiency of waste management, environment-friendly waste disposal and recycling systems, augmented waste transportation systems and AI–enabled waste prediction systems. The smart city waste management ecosystem can effectively function under a data-driven environment, which consists of sensor technologies, user interfaces, software solutions, IoT, seamless integration of data and digital platforms and automated control and surveillance of waste management infrastructures.

Machine learning, which is a branch of AI, facilitates use of data and algorithms, enables the computer to learn patterns from the input data and predict the required results, based on the patterns observed by the computer, during the time of learning done by the machines. Machine learning and AI enable the computer to analyse data, information, videos, images and other input information and provide results such as predictions, recommendations and other decision-oriented information for making decisions. In the case of smart city waste management, data derived from smart city waste monitoring sensor devices can be analysed based on various machine learning algorithms, and relevant requirements in terms of capacity and infrastructures can be predicted. Prediction of waste transportation and shipment, understanding the patterns of waste formations, prediction of waste recycling requirements and waste

DOI: 10.1201/9781003225317-5

disposal quantity predictions are some of the key outcomes due to incorporation of machine learning and AI within the waste management ecosystem.

5.2 LITERATURE REVIEW

The research work by Anh Khoa et al. (2020) proposed machine learning and a graph theory-based novel system for forecasting the likelihood of waste level in the garbage bins. This IoT-enabled system aimed to augment waste collection through the shortest path. Historical data was used to predict the probability of gathering the garbage in the real environment using a single microcontroller with an ultrasound sensor. This sensor was deployed to measure the filling level of the waste bin and transmit the details employing LoRa E32 technology. This research work was the first to suggest Dijkstra's algorithm for the best path for the waste collection and to use sigmoid function to predict the probability. The performance of the proposed system was tested in real time at Ton Duc Thang University. This system had the advantage that the sensor components can be changed quickly, making it suitable for practical applications. This is a low-cost system which collected data on time, optimizing the operations for employee use. Moreover, this system enhanced the profits and reduced the number of workers.

The research project by Gupta et al. (2019) suggested a smart dust bin using sensors, microcontrollers and Wi-Fi connections. The level of the waste determined by the ultrasonic sensor could indicate the quantity of waste. The sorting of items from the waste bins, such as reusable, recyclable and biodegradable items, was done by the weight and temperature sensors aided by metal detectors. Arduino Uno, based on ATMega328, was used, considering that it is of low cost and had minimal power usage. The main function of this microcontroller was to gather the data and transfer it to the internet. The data from the garbage bins was sent to the servers on the internet using Wi-Fi technology. MySQL software was used to organize the data collected. The statistical analysis was used to determine the filling rates of the garbage bins and decide the routes to be used by the team who are responsible for cleaning. The statistical analysis also helped in finding out the priority to be given for each location during the time of day. AI was deployed to enhance functionalities such as the route optimization based on the need and time of the day. This is required in places like offices and schools where cleaning should be done immediately than open fields which have less human interaction. AI-enabled optimization techniques focusing on the road traffic, time and driving distance helped to find the best routes, thereby reducing the cost and time. This could yield an extremely competent route to be taken by the drivers, augmented with GPS location services. This model is a very good waste collection scheme and is cost-effective.

Soni et al. (2019) evaluated six models of AI: Artificial Neural Network (ANN), Discrete Wavelet Theory ANN (DWT-ANN), Adaptive Neuro Fuzzy Inference System (ANFIS), Discrete Wavelet Theory ANFIS (DWT-ANFIS), Genetic Algorithm ANN (GA-ANN) and Genetic Algorithm ANFIS (GA-ANFIS) to assess the ability of each one in forecasting the quantity of waste being produced. Both primary and secondary data were used for this study. Primary data was collected through survey, and the secondary data was collected from various private, semi-governmental and governmental publications. Coefficient of determination (R^2), root mean square error (RMSE) and index of agreement (IA) values were determined for each model. A three-layered neural network with an input layer, a hidden layer and an output layer was used for the pure ANN model. Various combinations of delays and a distinct number of hidden nodes were also deployed to make several models. In the hybrid model DWT-ANN, discrete wavelet transformations were accomplished using Daubechies order 1(db1) mother wavelet to enhance the forecasting efficiency of the data. Further, this was fed into the ANN. For the GA-ANN model, genetic algorithm procedures were used to calculate the weights and the ideal biases for ANN as a substitute for the back-propagation technique. Various combinations of input nodes and hidden nodes were tested to obtain various models. The pure ANFIS model deployed triangle-shaped (trimf), trapezoid-shaped (trapmf) and Gaussian curve (gaussmf) function types for grid partitioning as the optimization method. Unfortunately, this model was not very successful in dealing with dynamic time-series data directly. So, a DWT-ANFIS model with pre-processing was deployed to improve the performance. This model used decomposition techniques using Daubechies, and noisy signals were also removed. Further, a GA-ANFIS algorithm was employed to augment biases and weights of ANFIS. The hybrid model of ANN and the genetic algorithm yielded the smallest RMSE, maximum IA and R^2 values. Consequently, this model has the highest accuracy among the six various models. The RMSE, IA and R^2 values of the six various models as per Soni et al. (2019) are given in Table 5.1.

A project by Awe et al. (2017) aimed to categorize the different types of waste into landfill, recyclable and paper using faster R-CNN. Faster R-CNN had a Region Proposal Network (RPN) which shared full image convolutional elements with the detection network. An RPN is a fully convolutional network that can predict both object bounds and objectiveness scores of every position. The RPN network was trained to produce high-quality region proposals that can be used by fast R-CNN for the purpose of detection. This project could obtain a mean average precision (mAP) of 0.683 on the overall classification of all the different types of the trash images.

A paper by Chowdhury and Chowdhury (2007) outlines a sensor and RFID (radio frequency identification)-based system for real-time waste management. The system had the features of automatic waste identification and weight and stolen bin

TABLE 5.1
RMSE Comparison Table

Model	RMSE	IA	R^2
Pure ANN	165.5	0.524	0.72
DWT-ANN	157	0.465	0.67
GA-ANN	95.7	0.864	0.87
Pure ANFIS	224.9	0.0982	0.36
DWT-ANFIS	155.4	0.682	0.73
GA-ANFIS	393.8	0.347	0.56

identification. The main components of this system are an RFID tag, a reader and an IT system for waste management. A smart device such as a smartphone or a PDA (personal digital assistant) is positioned inside the truck, which is used for waste collection. The antenna captures the radio waves from the reader apparatus connected to the smart device. This facilitates the chip to transfer its unique identification to the reading device. Thus, the bin can be identified remotely. The smart device attached to truck is very low weight, and the device is connected to the database server using a wireless network. This avoids the manual input of data and increases the efficiency and the accuracy of the system. The smart device is capable of converting the radio waves into digital signals. When the waste is loaded automatically into the truck using the robotic lifting arms, the weight of each bin is measured. This weight will be mapped with bin ID to charge each household for the waste disposal, and these details will be temporarily stored in the smart device.

Hong et al. (2014) suggested an IoT-enabled Smart Garbage System (SGS) with the aim to reduce the food waste. It is suggested for garbage bins with batteries, which are capable of exchanging data employing wireless mesh networks. This was made possible using a router and a server to collect and analyse the data for providing different services. This system had two domains: administration and service. The administration domain had all the details received from smart garbage bins, which are analysed and processed. The service domain had the information of the residents who threw the food waste in the smart bins. The service domain collects these details and transfers it to the administration domain. This system operated for one year in a Korean city and was found to reduce the food waste by average 33%.

5.3 STUDY OF EXISTING SYSTEMS

Modern smart city waste management ecosystems begin with facilitating sensor-based smart waste bins, which consists of measuring systems to monitor the fill levels of the bins. These smart bin operations can be enabled to use solar power and ensure that sustainability is achieved. Further, such systems also help to operate in remote areas, where solar energy is the only source of power. Fill level sensors with automatic signal generation, messaging to apps and data analysis systems can be incorporated to inform smart city administrators or municipality about information related to the filling of bins. Smart waste bins also can be incorporated with a compactor or a presser, which will facilitate pressing systems within the bins and increase the filing volumes of the bins. Modern smart city waste management empowers the citizen by providing more optimized and efficient waste management and helps to improve the environment in the city. The system also evaluates the bin infrastructure in the city in terms of efficiency, cost and density of the waste. Sensoneo is one of the smart city waste management systems that provide the solutions for optimization of waste bin capacity, location, types and waste frequency collections. These systems also provide information pertaining to traffic congestion in the city related to regular waste collection routes, collection of waste on demand and reflection of the actual needs for waste collection. The system also provides citizens with a free mobile app, enabling them to find the nearest available empty bins, directions to locations and single click notification to the operators. Monitoring of underground waste collections, digitization and capturing of waste management transportation routes and automating computing and monitoring activities also enhance smart city waste management systems. Integration of data analytics and machine learning helps the optimum utilization of waste management inventories, waste management infrastructures and waste management activities.

Smart city waste management solutions empower smart cities to manage a large quantity of municipal waste. Industrial growth and large-scale urbanizations are the two main factors leading to such municipal wastes. The waste production increased over a period of time during last decades due to the modern lifestyle around the globe and urban migrations of population. Sustainable smart city waste management needs effective, efficient and sophisticated tools to monitor, manage bins with sensors, optimize collection routes and optimize the waste management capacity in urban environments. Following are some of the existing systems achieved by waste management global solution providers.

5.3.1 Smart City Waste Management Solution

Modern smart city waste management solutions (Figure 5.1) provide innovative approaches by enabling smart cities to manage their waste effectively and efficiently. Such solutions also drive improved quality of life for smart city citizens and lower their footprint on the environment. Lack of efficiency and transparency are the biggest challenges faced by the smart city citizens. To overcome such problems, some of the smart city solutions are able to achieve 97% accuracy on waste productions and 30%–63% efficiency on waste collection routes. Smart hardware and software for waste management also facilitate underground bin monitoring, bin monitoring tracking for collections and manhole fill level monitoring using smart sensors and applications. Smart city people engagement through mobile apps ensures constant interactions with citizens of the smart cities during their mobility. Integration of state-of-the-art technology in waste management also helps to reduce the cost of

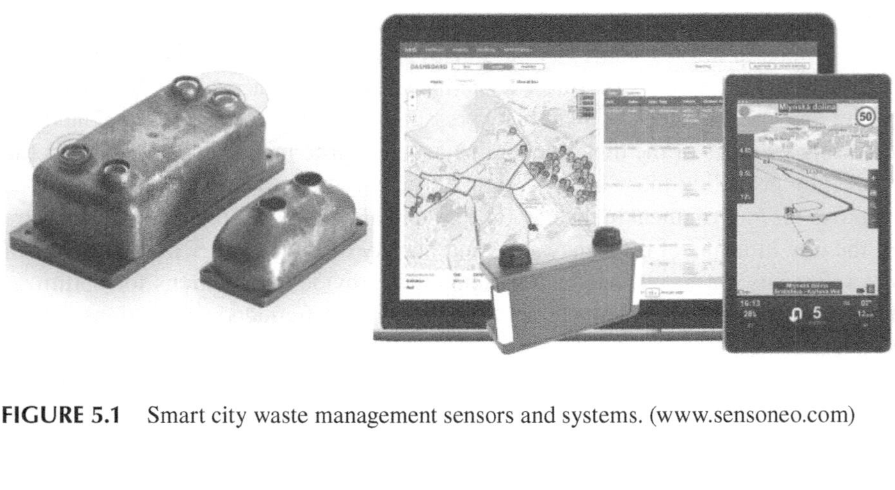

FIGURE 5.1 Smart city waste management sensors and systems. (www.sensoneo.com)

managing the wastes. A waste monitoring solution also monitors fill levels in the bins using smart sensors and ultrasound technology. These smart sensors are also able to monitor the fill levels of the bins or containers several times a day and send the data to smart management systems via IoT, thereby enabling optimization of waste collection routes, optimization of waste collection frequencies, vehicle loads and distributions of the bins. Such facilities help faster and more rapid data-driven decision-making capabilities of smart city waste management systems.

5.3.2 Smart City Asset Management Solution for Industrial Waste Management

Smart city asset management facilitates a clear understanding of managing the complex smart city bins, infrastructures, waste management stations, recycling plants and waste disposal equipment. Integration of radio frequency identification device (RFID) readers, smart waste management system components, RFID tags, RFID stickers and mobile apps ensure technology-enabled digitized asset management infrastructures and provides smart city-automated services to its citizens. Recording of waste management assets through the digital tagging of all bin containers, digitalization of waste management assets operation through cloud platforms, usage of RFID readers for automated service notifications and automatic updates on bin databases are some of the key features, which are put into service by modern waste management practitioners. Data-driven waste collection route planning and waste collection analysis enable the integration of machine learning and AI in the waste management solutions. Following are some of the highlights of smart city asset management solutions for industrial waste:

- Identify, evaluate, optimize and record all industrial wastes, infrastructures and assets.
- Simplify invoicing, communication and tracking of industrial waste management assets.
- RFID-enabled waste management service reporting and verification.
- Automated bin inventory management.
- Monitoring and tracking of industrial waste management infrastructure deployment.
- Automated failure reporting and automated service request on waste management assets.
- Dashboards with information about bin filling and utilization of industrial waste management assets.

5.4 AI-ENABLED MACHINE LEARNING FRAMEWORK FOR WASTE MANAGEMENT

Basic building blocks of AI–enabled machine learning frameworks for waste management facilitate the end-to-end integration of various predictions, clustering and classification techniques. Waste management data captured from

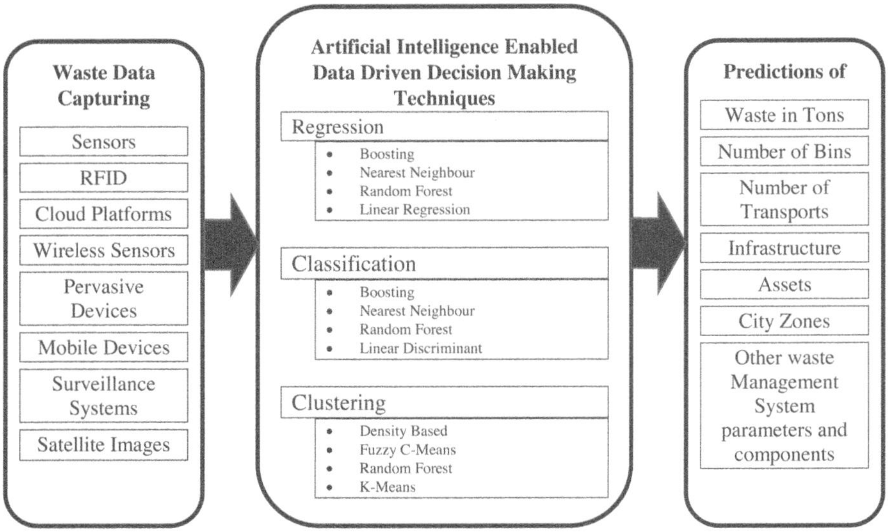

FIGURE 5.2 AI-enabled smart city waste management framework.

sensors, RFIDs, cloud platforms, wireless sensor networks, pervasive devices, mobile devices, surveillance systems and satellite transmissions are seamlessly integrated within the systems for analysis. These data normally generated from various smart city infrastructures such as smart bins, waste stations, waste transportation vehicles, waste recycling plants and land filling areas which normally fall within smart city geographical perimeters. Waste management activities such as waste generation, waste collections, waste recycling and waste filling are also an integral part of smart city waste management systems. The AI-enabled data-driven decision-making system, which uses various algorithms and models, can be made an integral part of waste management systems, and data generated or captured from various devices can be used for predicting and decision-making on waste management.

A basic AI-enabled waste management framework (Figure 5.2) facilitates the computer to process the data received from the devices using AI-enabled techniques such as regression techniques and classification techniques. The AI-enabled technology framework splits the input data, allows the machine or computer to learn from these data and then predict the results. The predicted results are compared with the original results, and best accuracy methods are used for predicting the required information.

Regression techniques facilitate prediction of waste generations in tons, prediction of number of bins needed, number of transports required for waste disposals, waste management infrastructures required and waste management assets required for managing the wastes. Classification and clustering techniques help to understand best routes, how the demographic area can be grouped, how the infrastructures can be classified or clustered for optimum usage of resources or any other classification and clustering of waste management parameters.

5.5 DESCRIPTION OF DATASET FOR DATA-DRIVEN DECISION-MAKING

Smart city data for simulations are assumed as per the requirements of the machine learning framework. A total of 10 variables, each consisting of 323 randomly assumed datasets, were used for smart city waste management data simulations. The following variables are used for analysing smart city waste management systems:

- Waste :Descriptions of smart city waste produced in tons.
- Bins :Number of smart bins used.
- Stations :Number of waste stations installed.
- Population :Smart city population in millions.
- Transportations :Number of transportations deployed for transporting wastes.
- Plants :Number of recycling plants installed.
- Area :Smart city areas in kilometres.
- Industries :Number of industries available within each city limit.
- Commercials :Number of commercial establishments producing the waste.
- Zones :City zones (1 = east, 2 = south, 3 = west, 4 = north).

Data structure rules and machine learning rules are followed while constructing the dataset, and algorithm requirements for processing data are studied. In Table 5.2, the tabular format of data is shown, which has been created based on assumptions and generated using random numbers.

During the input of data to the regression techniques, the dataset is grouped as per data variables and target variables. Target variables are the result values. Data variables are all other data columns. Further, splitting of dataset rows for training, validation and testing can be configured as per choices. Normally 60% data are used for training and 20% data used each for testing and validation purposes.

TABLE 5.2
Randomly Generated Waste Management Dataset

Serial No	Waste	Bin	Stations	Populations	Transportations	Plants	Area	Industries	Commercials	Zones
1	68,724	6641	68	65	691	34	661	55	678	2
2	57,934	5867	61	54	619	29	584	55	577	2
3	85,613	8665	89	85	866	44	850	15	877	1
4	83,916	8488	85	84	855	44	828	52	854	1
5	27,736	2786	33	27	354	12	289	19	306	4
–	–	–	–	–	–	–	–	–	–	–
–	–	–	–	–	–	–	–	–	–	–
–	–	–	–	–	–	–	–	–	–	–
321	9899	1082	17	11	189	3	181	67	169	4
322	58,308	5914	61	55	628	30	588	11	586	2
323	63,996	6349	65	61	669	32	636	88	652	2

TABLE 5.3
Exploratory Data Analysis

Variables	Count	Mean	Std	Min	25%	50%	75%	Max
Waste	323	51,059.7	28,291	385	28,943.5	50,100	75,781.5	99,531
Bins	323	5206.1	2812.8	100	2926.5	5261	7558.5	9976
Stations	323	54.7	25.6	10	35	57	74.5	100
Populations	323	50.1	27.3	1	27	49	71	100
Transports	323	562.1	247.6	104	363.5	571	771	999
Plants	323	25.6	14.7	1	12	25	39	50
Area	323	523	256.3	100	291	507	740.5	999
Industries	323	55.4	26.5	11	32	55	79	100
Commercials	323	538.5	258.2	108	311.5	537	759.5	999
Zones	323	2.6	1.1	1	2	3	3	4

5.5.1 Exploratory Data Analysis

Sample data exploratory analysis can be obtained by applying basic descriptive statistical tests on the dataset. Table 5.3 demonstrates descriptive statistics applied on the sample dataset. Information such as count, mean, standard deviation, percentiles and maximum value can be obtained for the analysis. Observe that there are 323 data (rows) and 10 variables (columns) available in the dataset.

Exploratory data analysis on a number of bins is carried out by generating histograms and density curves as shown in Figure 5.3. As a sample study, the distribution of bins in frequency spectrum is observed by capturing the skewness and kurtosis parameters. It was observed that the distributions are negatively skewed.

Data visualization on sample datasets can be obtained by generating box plots and relative histograms along with density plots. Comparison of mean values can be visualized from the box plots. A histogram with density plots provides the distribution of data over the frequency spectrum. The illustration in Figure 5.4 demonstrates

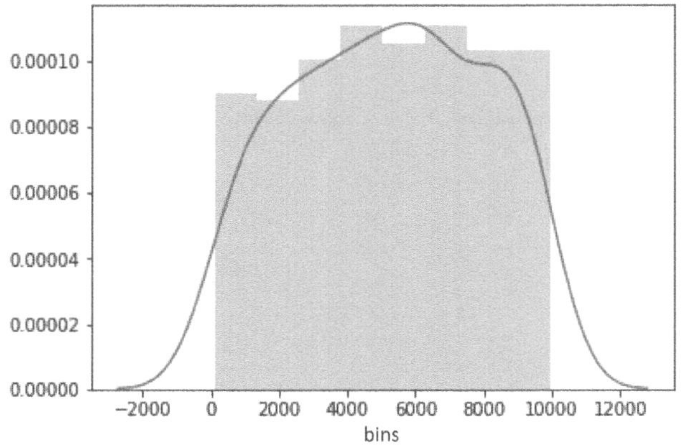

FIGURE 5.3 Density distribution curves.

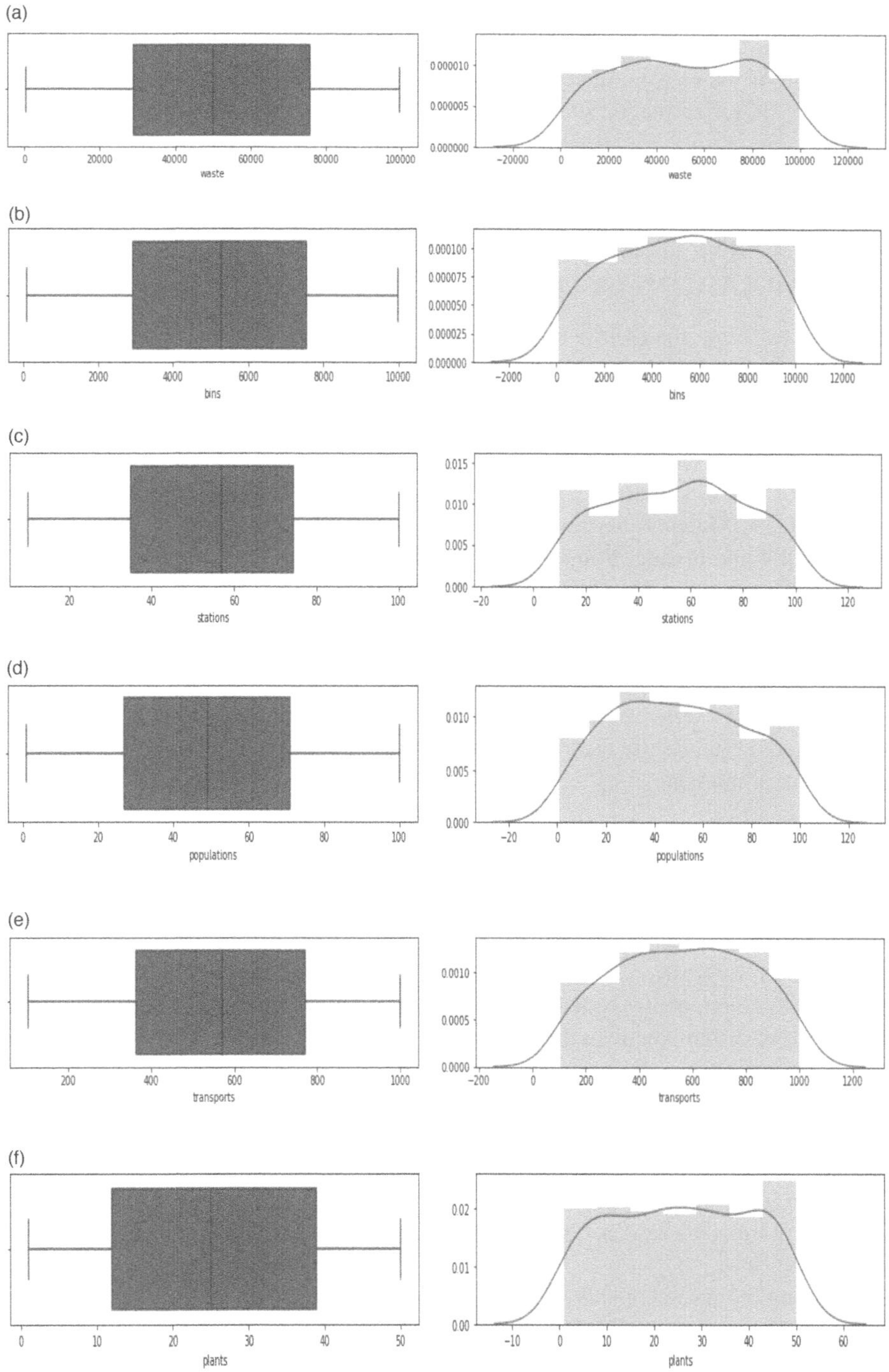

FIGURE 5.4 (a–f) Wastes, bins, stations, populations, transports, plants plots versus density distribution.

box plots and histogram plots with density curves of various waste data variables such as waste in tons, number of bins distributed, number of waste bin stations, population of the city, number of transportation provided for waste transportation, number of recycling plants, smart city area in kilometres, number of industries producing waste, number of commercial entities producing waste and classification of city zones based on demography.

5.6 MACHINE LEARNING ALGORITHMS AND SIMULATIONS OF DATASET

Machine learning algorithms convert the data into various data models. Machine learning algorithms are a set of rules that must be followed to obtain the required results. The results can be derived using regression analysis, classification analysis or clustering analysis. Each analysis may consist of several algorithms. Algorithms are a set of rules, which are followed to process the data in order to get the desired results. For machine learning, as a part of a dataset, results must also be included in the dataset. Machine learning processes the input dataset and splits the dataset into training and testing groups. Using the training dataset, a computer (machines) thoroughly learns the dataset, understands the patterns and produces the results in the form of predictions, clusters or classifications. Using the test dataset, a computer hides the existing results and predicts its own results, based on the learning that happened earlier by the machine. A comparison of predicted results with reference to actual results is carried out, and the accuracies of machine (computer) predictions are evaluated. Hence the accuracy determines the efficiency of machine learning algorithms or models.

5.6.1 Regression Algorithm for Predicting Waste Management Parameters

Prediction plays a significant role in AI, whenever data-driven decision-making is incorporated. Regression analysis helps one to predict smart city waste management parameters, components, systems, sub-systems, assets or any other smart city waste management system needs in the form of data values. For the assumed data, we can predict waste management infrastructures such as numbers of bins required, number of waste stations required, number of recycling plants required or prediction of any other requirements of waste management infrastructures. Four different types of regression analysis algorithm, such as boosting regression, K-nearest neighbours regression, random forest regression and regularized linear regression, are carried out on the same dataset. Each regression algorithm technique predicts the output as per its own rules.

Boosting regression consists of the iteration process along with strong predictor, which is constructed based on the previous weak predictor. Boosting is an ensemble tree approach, where each regression is created in the form of trees and subsequent trees are formed by adding all the previous trees. From the following data split parameters, we observe that the machine (computer) trained itself using 207 rows of data from the dataset. The machine also validated its learning using 52 datasets.

Further, the machine predicted 64 dataset results based on its learning. Whenever, MSE and RMSE of the results are the least, we can state that good prediction is possible.

From Table 5.4, we observe that test MSE is 0.003. If the MSE is low, best predictions are possible. Hence, we state that the best way of predicting the number of bins needed for the smart city waste management system is by boosting regression. The same steps can be followed to predict other information such as the requirement of smart city assets, infrastructures, transportation or prediction of zones and wastes.

From the data split parameters as shown in Figure 5.5, we observe that the machine (computer) trains itself using 207 rows of data from the dataset. The machine also validates its learning using 52 datasets. Further, the machine predicts 64 dataset results based on its learning.

From the evaluation matrices, as shown in Table 5.5, we observe that MSE is 0.003, which is the minimum.

If the MSE, RMSE and MAE are least, then good predictions are possible. Hence, using the boosting algorithm, best predictions can be achieved. From the previous table, we observe that R^2 value is 0.998. It means that 99.8% data can be fit on the

TABLE 5.4
Results of Boosting Regression

Boosting Regression							
Trees	**Shrinkage**	**Loss function**	**N (Train)**	**N (Validation)**	**N (Test)**	**Validation MSE**	**Test MSE**
80	0.100	Gaussian	207	52	64	0.002	0.003

Note: The model is optimized with respect to the *out-of-bag mean squared error*.

FIGURE 5.5 Training testing splits of dataset.

TABLE 5.5
Evaluation Metrics: Boosting Regression

Evaluation Metrics	
	Value
MSE	0.003
RMSE	0.055
MAE	0.038
MAPE	12.76%
R^2	0.998

TABLE 5.6
Relative Influence – Boosting Regression

Relative Influence

Attributes	Relative Influence
Waste in tons	80.438
Population-millions	13.663
No. of recycling plant	2.532
City zones	1.962
No. of waste stations	1.097
No. of commercial entities	0.223
No. of transportation	0.084
Area in km	0.000
No. of industries	0.000

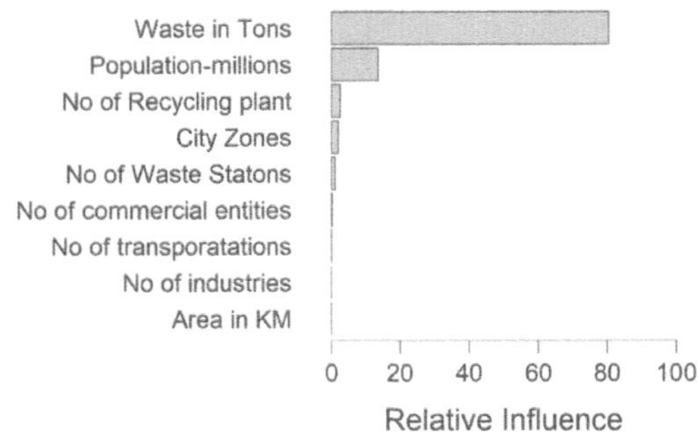

regression lines, which also proves that best predictions are possible through boosting regression.

The relative influence, provided in Table 5.6, demonstrates how much percentage each variable in the dataset influences predicting the outcomes.

From the previous table, we observe that waste in tons has the highest influence on predicting the smart waste bins for the smart city, followed by the population. The area of a smart city and the number of industries in the city have the least or zero influence on predicting the number of waste bins needed for the smart city.

It is also possible to visualize the predictive performance plot, which shows predicted values against their test set observations (Figure 5.6).

From the previous plot, we observe that data is closely fitted on the regression line; hence, we conclude that best fit and best prediction are possible.

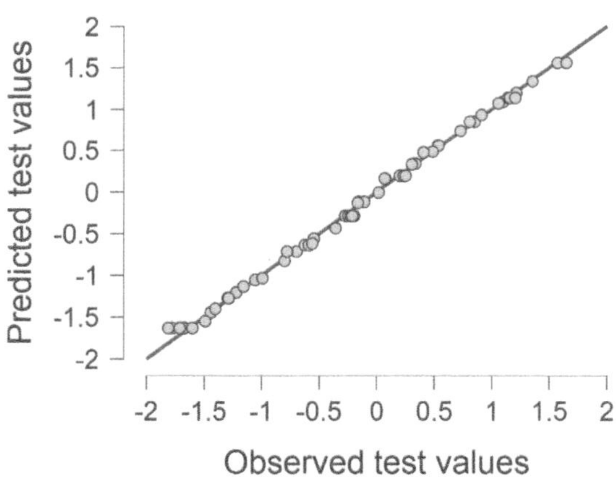

FIGURE 5.6 Predicted test values versus observed test values.

TABLE 5.7
Results of K-Nearest Neighbours Regression

K-Nearest Neighbours Regression							
Nearest Neighbours	Weights	Distance	N (Train)	N (Validation)	N (Test)	Validation MSE	Test MSE
2	Rectangular	Euclidean	207	52	64	0.007	0.004

Note: The model is optimized with respect to the *validation set mean squared error*.

K-nearest neighbours (KNN) regression predictions are done based on observing k number of predictors, which are similar to new observations. In KNN algorithm, the new value is predicted based on how closely it resembles the observed data, which is normally drawn from the training dataset during the training process and the testing dataset while predicting the outcomes using test dataset.

From Table 5.7, we observe that test MSE is 0.004. If the MSE is low, best predictions are possible. Hence, we state that the best way of predicting the number of bins needed for the smart city waste management system can also be done by K-nearest neighbour regression. The same steps can be followed to predict other information such as requirement of smart city assets, infrastructure, transportations or prediction of zones and wastes.

From Table 5.8 of evaluation matrices, we observe that MSE, which is 0, is the minimum. If the MSE, RMSE and MAE are least, good predictions are possible. Hence, using the K-nearest neighbours algorithm, best predictions can be achieved. From the table, we observe that R^2 values are 1. That means, 100% data can be fitted on the regression lines, which also proves that best predictions are possible through the K-nearest neighbour algorithm.

Table 5.9, depicting variables of importance, demonstrates how much percentage for each variable in the dataset is important for predicting the outcomes. From the table, we observe that a number of commercial entities have the highest influence on

TABLE 5.8
Evaluation Metrics Results of K-Nearest Neighbour Regression

Evaluation Metrics	
Performance Measure	Value
MSE	0
RMSE	0
MAE	0.012
MAPE	7.76%
R^2	1

TABLE 5.9
Variable of Importance – Boosting Regression

	Variable Importance	
	Mean Decrease in Accuracy	**Total Increase in Mode Purity**
No. of commercial entities	0.173	19.362
Waste in tons	0.174	18.227
Population (in millions)	0.169	15.470
Area (in km)	0.147	14.634
No. of recycling plant	0.118	13.268
No. of transportations	0.129	13.258
No. of waste stations	0.100	10.065
City zones	0.024	2.605
No. of industries	1.176e−5	0.004

predicting the smart waste bins for the smart city, followed by waste in tons. City zones and the number of industries in the smart city have the least and zero importance on predicting the number of waste bins needed for the smart city.

Random forest regression techniques use an ensemble of a large number of individual decision trees to create a set of integrated decision trees. The ensemble method combines multiple predictions to derive the single powerful predictions. The techniques can be applied to predict various waste management system outcomes such as prediction of smart city zones, bins, infrastructure and assets. By applying random forest regression techniques on the dataset, we can generate the predictions.

From Table 5.10, we observe that test MSE is 0. If the MSE is low, best predictions are possible. Hence, we state that the best way of predicting the number of bins needed for the smart city waste management system can be done by random forest regression. The same steps can be followed to predict other information such as requirement of assets, infrastructure, transportation or prediction of zones and wastes.

From Table 5.11, depicting evaluation matrices, we observe that MSE is 0, which is also the minimum. If the MSE, RMSE and MAE are least, then good predictions are possible. Hence, using random forest algorithm, best predictions can be achieved. From the result table, we observe that R^2 values are 1. That means, 100% data can

TABLE 5.10
Results of Random Forest Regression

Random Forest Regression							
Trees	**Predictors per Split**	**N (Train)**	**N (Validation)**	**N (Test)**	**Validation MSE**	**Test MSE**	**OOB Error**
100	3	207	57	64	3.566e−4	3.214e−4	3.475e−4

Note: The model is optimized with respect to the *out-of-bag mean squared error.*

TABLE 5.11
Evaluation Metrics: Random Forest Regression

Evaluation Metrics	
	Value
MSE	0
RMSE	0
MAE	0.014
MAPE	68.88%
R^2	1

be fitted on the regression lines, which also proves that best predictions are possible through random forest algorithm.

Table 5.12, describing variables of importance, demonstrates how much percentage each variable should have in the dataset to be important in predicting the outcomes. From the table, we observe that the number of commercial entities has highest influence on predicting the smart waste bins for the smart city, followed by waste in tons. City zones and number of industries in the smart city have the least or zero importance on predicting the number of waste bins needed for the smart city.

Regularized linear regression analysis provides a general linear relationship between the input variables and provides the predictions of target variables for the input data. The prediction of the regularized linear model is based on the weighed sum of averages. Further, regularized linear regression is an extension of linear regression, in which coefficients are shrunk to zero.

From Table 5.13, we observe that test MSE is 0.001. If the MSE is low, then best predictions are possible. Hence, we state that the best way of predicting the number

TABLE 5.12
Random Forest Variables of Importance Information

Variable Importance		
	Mean Decrease in Accuracy	**Total Increase in Node Purity**
No. of commercial entities	0.186	18.179
Waste in tons	0.175	16.905
No. of transportation	0.163	15.657
Area in KM	0.149	15.156
No. of waste stations	0.143	13.856
No. of recycling plant	0.141	13.840
Population-millions	0.119	11.863
City zones	0.005	0.511
No. of industries	6.702e−6	0.006

TABLE 5.13
Results of Regularized Liner Regressions

Regularized Linear Regression						
Penalty	**λ**	**N (Train)**	**N (Validation)**	**N (Test)**	**Validation MSE**	**Test MSE**
L1 (Lasso)	0.007	207	52	64	9.751e−4	0.001

Note: The model is optimized with respect to the *validation set mean squared error.*

of bins needed for the smart city waste management system can be done by regularized linear regression. The same steps can be followed to predict other information such as requirement of assets, infrastructure, transportation or prediction of zones and wastes.

From Table 5.14, depicting evaluation matrices, we observe that MSE is 0.001, which is the minimum. If the MSE, RMSE and MAE are least, then good predictions are possible. Hence, using the random forest algorithm, best predictions can be achieved. From the result table, we observe that R^2 value is 0.999. That means, 99.9% data can be fitted on the regression lines, which also proves that best predictions are possible through generalized linear regression.

5.6.2 Classification Algorithm for Predicting Waste Management Parameters

Classification techniques are used to predict class labels for the given set of input data. In an AI-enabled decision-making process, classification facilitates predicting categories of various devices used for managing wastes, types of wastes, locations, waste management infrastructures and waste management assets. Classification categorizes the input data into classes, which helps to predict the category of the input. In machine learning and AI, many algorithms are used for classification problems. However, we are discussing four algorithms, which are boosting classification,

TABLE 5.14
Evaluation Metrics of Regularized Linear Regression

Evaluation Metrics	
	Value
MSE	0.001
RMSE	0.032
MAE	0.027
MAPE	5.67%
R^2	0.999

TABLE 5.15
Results of Bosting Classification

Boosting Classification						
Trees	Shrinkage	N (Train)	N (Validation)	N (Test)	Validation Accuracy	Test Accuracy
71	0.100	207	52	64	1.000	0.984

Note: The model is optimized with respect to the *out-of-bag accuracy*.

K-nearest neighbour classification, linear discriminant classification and random forest classification.

Boosting classification (Table 5.15) predicts the final outcomes based on sequential and successive weak predictors using decision tree ensemble techniques. Boosting classification processes the input data using multiple iterations. During every successive iteration, new strong predictors are created based on the previous weak predictors until final prediction is achieved. The evaluation of classification algorithm is done based on accuracy achieved. If the test accuracy is 1, it is interpreted as 100% accuracy achieved and the selected algorithm is best suited for the predictions. For the classification purposes, we have selected zones as predictors. Hence for the given input, which consists of nine variable (column) waste management data, a classification algorithm predicts the location of the smart city (1 = east zone of the smart city, 2 = south, 3 = west and 4 = north).

From Table 5.15, we observe that, for the smart city waste management test dataset, predictions of smart city zones with 98.4% accuracy can be achieved using boosting classification. Table 5.16 demonstrates the confusion matrix, where the observed values versus predicted values are compared for the given dataset.

From Table 5.16, we observe that predictions were carried out on 64 datasets, and only one prediction was found wrong. For the zone 4 data, the computer predicted it as zone 3. Hence, 98.4% accuracy was achieved.

TABLE 5.16
Confusion Matrix of Boosting Classification

		Confusion Matrix Predicted			
		Zone 1	Zone 2	Zone 3	Zone 4
Observed	Zone 1	13	0	0	0
	Zone 2	0	16	0	0
	Zone 3	0	0	20	0
	Zone 4	0	0	1	14

TABLE 5.17
Results of the K-Nearest Neighbour Classification

K-Nearest Neighbour Classification							
Nearest Neighbours	Weights	Distance	N (Train)	N (Validation)	N (Test)	Validation Accuracy	Test Accuracy
7	Rectangular	Euclidean	207	52	64	0.962	0.906

Note: The model is optimized with respect to the *validation set accuracy.*

K-nearest neighbour classification (Table 5.17) observes k number of similar data, which are available nearer to the selected data and predict the outcomes based on similarities found in the nearest data. For the input waste management data, the K-nearest neighbour algorithm predicts the location of the smart city.

From Table 5.17, we observe that, for the smart city waste management test dataset, predictions of smart city zones with 90.6% accuracy can be achieved using K-nearest neighbour classification. Table 5.18 demonstrates the confusion matrix, where the observed values versus predicted values are compared for the given dataset.

From Table 5.18, we observe that predictions were carried out on 64 datasets and only two predictions were found wrong. For the zone 2 data, the computer wrongly predicted it as zone 1. Similarly for the zone 4 data, the computer wrongly predicted it as zone 3. Hence, 90.6% accuracy was achieved.

Linear discriminant analysis (LDA) classification (Table 5.19) predicts the outcome using dimensionality reduction techniques. LDA classification projects the features available in the higher dimension space onto lower dimension space and thereby identifies the patterns using less resources and costs. For the smart city waste management data, the LDA algorithm can predict the locations of a smart city.

From Table 5.19, we observe that, for the smart city waste management test dataset, predictions of smart city zones with 98.4% accuracy can be achieved using the LDA classification. Table 5.20 demonstrates the confusion matrix, where the observed values versus predicted values are compared for the given dataset.

TABLE 5.18
Confusion Matrix of K-Nearest Neighbour Classification

		Confusion Matrix			
		Predicted			
		Zone 1	Zone 2	Zone 3	Zone 4
Observed	Zone 1	12	0	0	0
	Zone 2	1	12	0	0
	Zone 3	0	0	20	0
	Zone 4	0	0	1	14

TABLE 5.19
Results of Linear Discriminant Classifications

Linear Discriminant Classification				
Linear Discriminants	**Method**	**N (Train)**	**N (Test)**	**Test Accuracy**
3	Moment	259	64	0.984

TABLE 5.20
Confusion Matrix of Linear Discriminant Classification

		Confusion Matrix			
		Predicted			
		Zone 1	**Zone 2**	**Zone 3**	**Zone 4**
Observed	Zone 1	11	0	0	0
	Zone 2	0	20	0	0
	Zone 3	0	0	20	0
	Zone 4	0	0	1	12

TABLE 5.21
Results of Random Forest Classification

Random Forest Classification							
Trees	**Predictors per Split**	**N (Train)**	**N (Validation)**	**N (Test)**	**Validation Accuracy**	**Test Accuracy**	**OOB Accuracy**
1	3	207	52	64	0.962	0.984	1.000

Note: The model is optimized with respect to the *out-of-bag accuracy*.

From Table 5.20, we observe that predictions were carried out on 64 datasets, and only one prediction was found wrong. For the zone 4 data, the computer predicted it as zone 3. Hence, 98.4% accuracy was achieved.

Random forest classification (Table 5.21) builds multiple classification decision trees and merge all of them to get one more accurate classification decision tree. During the process, the random forest algorithm searches the best feature among the set of features it retrieved from the dataset. For the smart city waste management data, a random forest classification algorithm can predict the location of the smart city.

From Table 5.21, we observe that, for the smart city waste management test dataset, predictions of smart city zones with 100% accuracy can be achieved using a random forest classification. Table 5.22 demonstrates the confusion matrix, where the observed values versus predicted values are compared for the given dataset.

TABLE 5.22
Confusion Matrix of Random Forest Classification

		Confusion Matrix Predicted			
		Zone 1	**Zone 2**	**Zone 3**	**Zone 4**
Observed	Zone 1	16	0	0	0
	Zone 2	0	13	0	0
	Zone 3	0	0	20	0
	Zone 4	0	0	0	14

From Table 5.22, we observe that predictions were carried out on 64 datasets, and the predictions were found correct. Hence, 100% accuracy was achieved.

5.6.3 Clustering Algorithm for Predicting Waste Management Parameters

The clustering technique uses labelled input data and processes to find meaningful information, features and generate groups. A cluster divides the dataset into several groups by observing similarities in the data points. There are many different clustering algorithms such as fuzzy C-means clustering, hierarchical clustering, K-means clustering and random forest clustering, which can also be used for processing waste management dataset.

Fuzzy C-means clusters (Table 5.23) process the results based on distances between the data and the cluster centre. The nearer to the centre point, the more similarities between the data. Hence, as per the fuzzy C-means algorithms, these similar data then form a cluster. Similarly, a number of clusters can be created by dividing the population or the dataset. In a waste management dataset, clusters on smart city locations based on specific key aspects, clusters of waste stations based on waste bins, creation of recycling plants based on specific clusters or any other group formation based on waste data information are used.

TABLE 5.23
Results of Fuzzy C-Means Clustering

Fuzzy C-Means Clustering					
Clusters	**N**	**R^2**	**AIC**	**BIC**	**Silhouette**
9	32.3	0.926	382.060	688.050	0.350

Note: The model is optimized with respect to the *BIC* value.

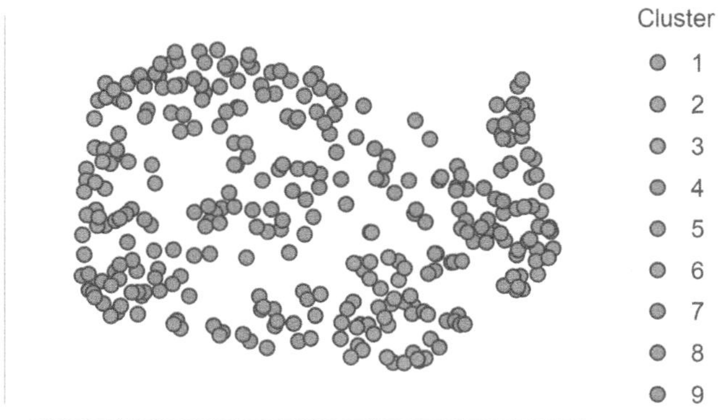

FIGURE 5.7 t-SNE plot on cluster formation.

From Table 5.23, we observe that the fuzzy C-means algorithm creates nine clusters with a silhouette score of 0.350. The higher the silhouette score, the better formation of clusters. Hence, the formation of nine good clusters is possible, without overlapping of the data.

Data visualization can be achieved by plotting a t-SNE plot. The t-SNE plot (Figure 5.7) demonstrates the visualization of high-dimensional data in low-dimensional space by establishing relative distances between data observations.

5.7 RESULT ANALYSIS AND CONCLUSION

Comparison of different regression techniques reveals, for the given waste management dataset, which one is the best algorithm model suitable for predicting various waste management information. Table 5.24 provides the comparative analysis on four regression techniques, which has been used for the prediction of waste bins, as a part of an AI-enabled, data-driven, decision-making framework.

From Table 5.24, comparing evaluation parameters such as MSE, RMSE, MAE, MAPE and R^2, we observe that all the algorithms were found to be good for

TABLE 5.24
Evaluation Metrics of Regression Techniques

	Evaluation Metrics			
Parameters	**Boosting**	**K-Nearest Neighbour**	**Random Forest**	**Regularized Linear Regression**
MSE	0.003	0	0	0.001
RMSE	0.055	0	0	0.032
MAE	0.038	0.012	0.014	0.027
MAPE	12.76%	7.76%	68.88%	5.67%
R^2	0.998	1	1	0.999

TABLE 5.25
Evaluation Metrics of Classification Techniques

	Evaluation Metrics			
Parameters	**Boosting Classification**	**K-Nearest Neighbour Classification**	**Linear Discriminant Analysis Classification**	**Random Forest Classification**
Test accuracy	0.984	0.906	0.984	1

predicting waste management system requirements. However, the K-nearest neighbour model emerges as the best fit among all the four models used. Hence, in the case of regression, techniques are incorporated, which is recommended to use the K-nearest neighbour algorithm.

Comparison of different classification techniques reveals, for the given waste management dataset, which one is the best algorithm model suitable for classifying or categorizing the waste management infrastructures. Table 5.25 provides the comparative analysis on four classification techniques, which is used for the classification of smart city locations.

From Table 5.25, comparing evaluation parameters test accuracy, we observe that the random forest algorithm is best suited for solving classification problems with reference to classification of locations. The same steps can be followed for creating other classifications with reference to waste management systems.

Comparison of different clustering techniques reveals, for the given waste management dataset, which one is the best algorithm model suitable for clustering or grouping of waste management information. Here the groups or clusters are formed by the algorithm based on its own rules. Table 5.26 demonstrates the comparison of clustering algorithms.

In the clustering method, the higher the silhouette score is treated, the better the model is obtained. From Table 5.26, by comparing evaluation parameters such as the number of clusters and silhouette score, we observe that K-means clustering (high silhouette score) is best suited for solving clustering problems. The same steps can be followed for creating other clusters with reference to various other data related to waste management systems.

TABLE 5.26
Evaluation Metrics of Clustering Techniques

	Evaluation Metrics			
Parameters	**Fuzzy C-Means Clustering**	**Hierarchical Clustering**	**K-Means Clustering**	**Random Forest Clustering**
Number of clusters	9	7	7	6
Silhouette score	0.350	0.350	0.380	0.260

REFERENCES

Anh Khoa, T., Phuc, C. H., Lam, P. D., Nhu, L. M. B., Trong, N. M., Phuong, N. T. H., Dung, N. van, Tan-Y, N., Nguyen, H. N., & Duc, D. N. M. (2020). Waste management system using IoT-based machine learning in university. *Wireless Communications and Mobile Computing*, 2020. https://doi.org/10.1155/2020/6138637

Awe, O., Mengistu, R., & Sreedhar, V. (2017). *Final Report: Smart Trash Net: Waste Localization and Classification.*

Chowdhury, B., & Chowdhury, M. U. (2007). *This Is the Published Version: RFID-based Real-Time Smart Waste Management System.* http://hdl.handle.net/10536/DRO/DU:30008105

Gupta, P. K., Shree, V., Hiremath, L., & Rajendran, S. (2019). The use of modern technology in smart waste management and recycling: Artificial intelligence and machine learning. In *Studies in Computational Intelligence* (Vol. 823, pp. 173–188). Springer Verlag. https://doi.org/10.1007/978-3-030-12500-4_11

Hong, I., Park, S., Lee, B., Lee, J., Jeong, D., & Park, S. (2014). IoT-based smart garbage system for efficient food waste management. *Scientific World Journal*, 2014. https://doi.org/10.1155/2014/646953

Soni, U., Roy, A., Verma, A., & Jain, V. (2019). Forecasting municipal solid waste generation using artificial intelligence models—A case study in India. *SN Applied Sciences*, *1*(2). https://doi.org/10.1007/s42452-018-0157-x

Udupi, P. K. (2023). Practical machine learning: No-coding, easy and fast track using JASP – A free and open source software. amazon.com

6 Blueprint of IoT for Smart Cities
An Emerging Markets Perspective

Riju Ghosh[1], *Amol S. Dhaigude*[2], *and Anudeep-Sri Bathina*[3]
[1]Amity Business School, Noida, Uttar Pradesh, India
[2]S. P. Jain Institute of Management and Research (SPJIMR), Mumbai, Maharashtra, India
[3]University of Massachusetts Dartmouth, Dartmouth, Massachusetts

6.1 INTRODUCTION

Today, whenever one thinks of quality urban living, "smart city" is the term which rings a bell to the mind. With the advent of technology and the onset of digital transformation across all walks of life, information and communication technology (ICT) have become fundamental building blocks of cities (Caragliu, Del Bo and Nijkamp, 2011; Batty et al., 2012; Chourabi et al., 2012; Ismagilova et al., 2019; Khan et al., 2022). Internet of Things (IoT), the technology and process of connecting things apart from the common networking devices to the internet, has gathered an ever-increasing momentum. IoT is not just a buzzword anymore but an essential technology (e.g., Khan and Salah, 2018). Smart cities are thriving all across the globe with governments as well as private players pushing to improve the quality of life in urban spaces. The urban areas are fast expanding and thus having core infrastructure resources, such as water, electricity, and transportation, enveloped in smart solutions has become a dire need (Letaifa, 2015). In this chapter, we present IoT as the major solution provider and foundational technology for smart cities and propose a framework that can help formulate the digital layout of the smart cities.

It is very important to realize the fact that IoT and its related technology are imperative when it comes to smart cities (e.g., Chatterjee, Kar and Gupta, 2018; Saravanan, Julie and Robinson, 2019). As a matter of fact, if there are no smart solutions, then there is no smart city. Hence, connected devices and their underlying architecture are vital to understand the use cases which are pertinent to smart cities (Lea and Blackstock, 2014). Once one acknowledges IoT and then delves into the realms of smart cities then the design, development, and their evaluation all become simplified (Rajab and Cinkelr, 2018). Even the frameworks and assessment matrices

DOI: 10.1201/9781003225317-6

we discuss in this chapter will showcase why IoT is so crucial in the context of smart cities. Thus, this article addresses questions such as: Can there be a smart city without IoT? What makes IoT the central component of smart cities? How is IoT shaping up the smart cities? And what impacts do IoT-based solutions have on the effectiveness of a smart city?

In this chapter, we approach the story of IoT and its relevance in smart cities in a layered manner. First, we understand what IoT is and its architecture from different perspectives. Second, we encapsulate what a smart city is and what its components are. Then, we understand how and where IoT plays a crucial role in the planning and development of a smart city. Frameworks and assessment matrices help us understand the effectiveness of the IoT-based smart solutions for a smart city. Finally, we discuss two key frameworks and assessment matrices along with common use cases in a smart city which showcase the IoT technology that is being used. Throughout the chapter, we have taken examples and use cases wherever needed to illustrate the idea. We also discuss how India, an emerging economy, is building and sustaining smart cities. The very real-world challenges of IoT implementation in the smart city diaspora are also discussed in this chapter.

By delving into the various practical frameworks and use cases of smart cities, we find smart cities are largely dependent on IoT-based solutions. A smart city's planning and development phase relies hugely upon smart solutions. At the same time, the effectiveness and impact are related mainly to IoT and connected technologies. Thus, the more efforts made towards digitizing the urban area, the more effective and resilient the smart city is. The findings of this article will help academia and practitioners in multiple ways. First, the decision-makers will realize the importance of IoT for building and maintaining smart city. This will encourage them to invest in IoT to achieve smart city goals. The framework discussed by us will help scholars and practitioners progress in the field of research and practice.

The rest of the chapter has been divided into multiple sections. Section 6.2 deals with conceptual background/literature review. Section 6.3 talks about IoT architecture from a business/enterprise perspective. Section 6.3 speaks about smart cities, whereas Section 6.4 deals with the framework and assessment matrices of smart cities. Section 6.5 talks about the Indian context, and in Section 6.6, IoT highlights smart cities' common use cases. Section 6.7 highlights the conclusion, discussion, and implications. Finally, Section 6.8 deals with limitations and directions for future research.

6.2 CONCEPTUAL BACKGROUND

6.2.1 Smart Cities and IoT

The term "smart city" signifies combination and assimilation between the ICT and urban areas (Sterbenz, 2017). The IoT is one of the most important aspects in the successful implementation of a smart city (Jin et al., 2014). For instance, the data generated by electric bikes or cars is shared and stored in a centralized location to give citizens highly customized offerings (Zanella et al., 2014; Samuel, 2016).

The smart city, like the notion of a smart house, is an emerging market and a vital facet of upcoming infrastructure. As a smart city seeks to use energy and electricity effectively, IoT technology is even more important, therefore offering a pleasant and economically sensible architecture. Consequently, once a smart city's fundamental infrastructure is in place, numerous services that need the usage of various forms of data collected in daily life may be incorporated. Therefore, numerous IoT-supported services may benefit residents residing in sustainable and pleasant surroundings in the smart city.

Only some major sectors are collaborating to generate a responsible urban environment, considering the social and technological concerns of future smart city notions. For example, the energy sector (smart grid), healthcare sector (digital records), security sector (voice and video communications), and automobile sector (remote-control services) are taking initiatives to collaborate for the development of the smart city. Furthermore, previous research outlined that the major technological features of smart cities and IoT technologies may be deemed a precondition (Ash and Chandrasekaran, 2017).

6.3 WHAT IS IoT?

The Internet of Things (IoT) is a buzzword which has grasped industry and academia alike. The use of IoT to its fullest potential remains the horizon of a vast ocean which is being explored at a good pace by mankind. The term IoT was first used in 1999 by Kevin Ashton, the co-founder of MIT's (Massachusetts Institute of Technology) Auto-ID Centre (a pioneering research group in the field of RFID-radio frequency identification), when he was working for Proctor and Gamble. One of the earliest known use cases of IoT was a Coca-Cola vending machine, which was used at Carnegie Mellon University. It could communicate whether the drinks which were newly loaded were cold or not; also, it had the power to describe its inventory over ARPANET (Advanced Research Projects Agency Network).

We will define IoT in layman's terms before we delve further into the nitty gritty of IoT and its underlying technologies. IoT is the process and technology of connecting and communicating with things which usually do not have internet connectivity. For instance, a regular refrigerator was used for preserving the food items, but with IoT, it evolves into a smart fridge, which now has access to the internet and communicates with you the list of items it has over a mobile app. Furthermore, it can also order food items such as butter and juice, which you replenish time and again from various e-commerce websites, thus reducing the human efforts. The by-product of this process is data that is stored and processed using various machine learning algorithms over the cloud, which can help the companies in various business scenarios better understand the customer's replenishment cycle and market the right product to the customer. Figure 6.1 showcases the use case diagram of a smart refrigerator.

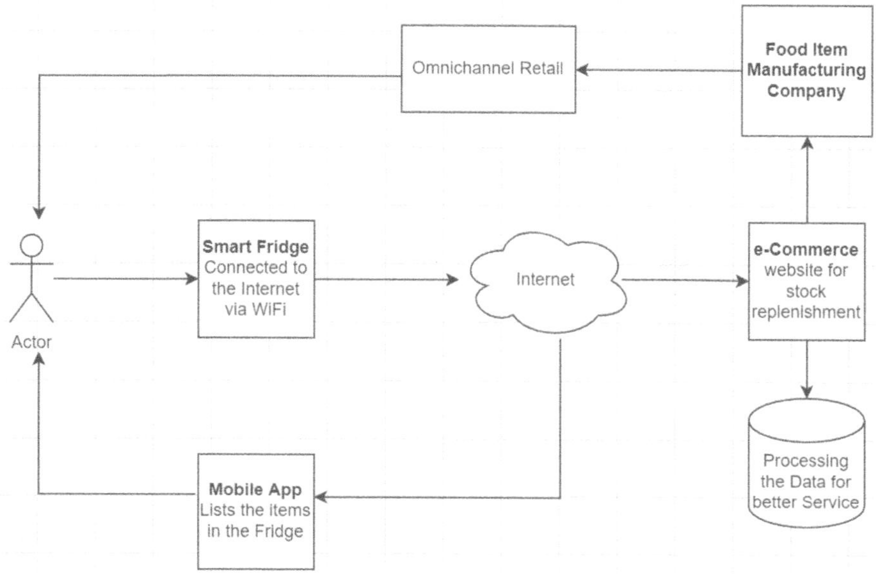

FIGURE 6.1 Use case diagram of a smart refrigerator.

6.4 IoT ARCHITECTURE

When the question arises for one common architecture for IoT systems, a single consensus has not been reached so far; there are a myriad of architectures depending on the use case of the system being developed. Here, we would like to discuss some of the commonly used ones and will attempt to capture various perspectives related to the designing of an IoT system.

6.4.1 Developers' and System Designers' Perspectives

The IoT system is one of the most crucial stages as it encompasses the very genesis of the use cases and the problems being solved or the processes being optimized. Therefore, having an understanding and insight into the structure of development becomes imperative for organizations or individuals which/who wish to deploy IoT systems. The most commonly used architecture of four layers is shown in Figure 6.2 and explained as follows:

1. **Things/physical layer:** This layer is a real physical layer, directly connected to the real-world objects by virtue of sensors and actuators connected to a microcontroller board. This layer is responsible for collecting the data by sensing it from the outside world. The collected data is formatted into the required structure and then sent to the upper layer by virtue of communication technologies, such as Bluetooth, zig-bee, LoRa (Low-power Wide Area Network), and WiFi, which are interfaced to the given microcontroller unit.

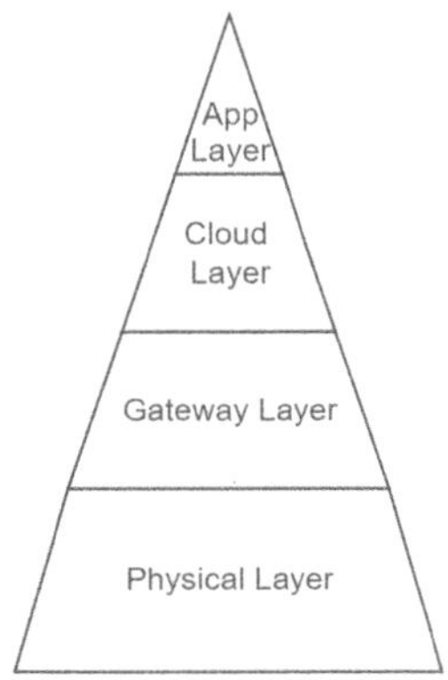

FIGURE 6.2 IoT architecture from design and development perspectives.

2. **Gateway layer:** This layer, also referred to as the network layer, acts as a bridge between the things layer and the cloud layer. Once the data is collected in the microcontroller units, it is sent to this layer which is basically a microcomputer with some storage capabilities and proper two-way communication and access to the internet. Now, this data is processed to an extent to change it into the form of suitable packets to be sent to the cloud platform by virtue of protocols, such as MQTT (Message Queueing Telemetry and Transport) and CoAP (Constrained Application Protocol), which are used specifically for IoT or by using regular protocols such as TCP/IP (Transmission Control Protocol/Internet Protocol) and UDP (User Datagram Protocol).
3. **Cloud/data-processing layer:** This is the layer which consists of servers kept in data centres, and the real processing of the data happens here. The data can be exposed to various artificial intelligence and machine learning algorithms based on the use case if it is a regression problem or a classification problem. The data is converted into useful information after computing in this very layer, and the storage of large volumes of data also happens here. This layer is where it is decided to what to make of the data and what useful action or decision is to be taken based on the given use case at hand.
4. **Application layer:** This layer is like a human-computer interface between the cloud layer and the end-user or -users of the data; here, the processed data and its related information is displayed to the user by the virtue of web application and/or mobile application.

6.5 ILLUSTRATIVE USE CASE OF IoT FROM A DEVELOPER'S PERSPECTIVE: SMART FARM

We will understand the development architecture of IoT discussed in a better manner with the help of an illustrative example of a use case. We will discuss elements in a smart farm for the IoT system.

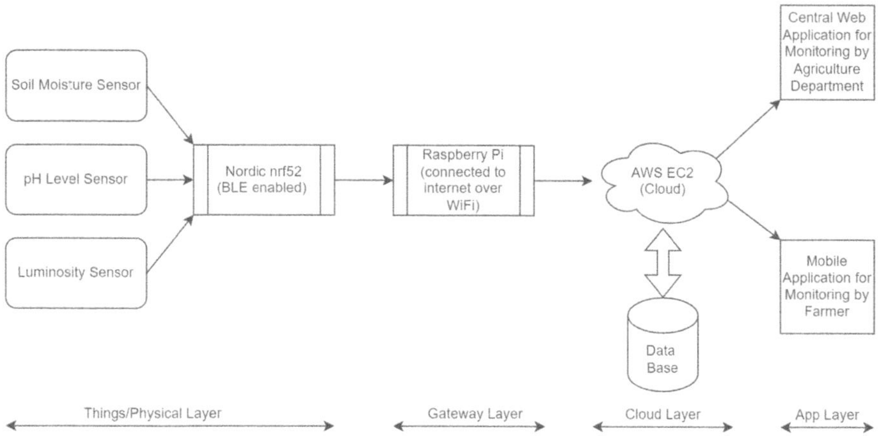

FIGURE 6.3 Block diagram of a smart farm use case of an IoT system.

Figure 6.3 shows the use case of a smart farm; here we have used the IoT architecture from the perspective of a developer/designer. Let's understand the components of the system individually.

1. **Microcontroller board with sensors:** In this case, the microcontroller board used is based on Nordic Semiconductor's nRF52 processor and board; the board uses BLE (Bluetooth Low Energy) as a communication medium. The sensors are directly connected to the board, and the data is collected from the soil comprising soil moisture level, pH level, and the incident light level or luminosity. This data so collected is sent over BLE to the Raspberry Pi for further use.
2. **Raspberry Pi gateway:** Raspberry Pi is a microcomputer which receives the data from nRF52 over BLE and then further sends the data to the cloud platform by virtue of an internet connection over WiFi (wireless fidelity). The point to be noted here is that multiple nRF52 boards can be connected to one Raspberry Pi; thus, one gateway can support multiple instances of a physical layer. Thus, acres of farm land can be served by using a combination of one gateway with multiple edge devices which are spread across small strips of farm land.
3. **Cloud layer:** The data from individual edge devices is aggregated in Raspberry Pi and sent to the AWS cloud hosting an EC2 (elastic cloud 2) instance, which is like a web server and database to store the data as well to perform data analytics over it; the quality of the soil and its needs for growing the crops can be assessed here very well and all necessary actions can be taken.
4. **Application layer:** A web-based app is provided to central authority to monitor the soil conditions of a given region, whereas the farmer can monitor his/her own farms' soil quality by using the mobile app; both these apps use the data in the cloud achieved from edge devices.

6.5.1 IoT Architecture from a Business/Enterprise Perspective

Businesses have been aggressively adapting IoT systems into their solutions and operations; their perspective is similar to that of a developer but a little less granular as compared to that of an IoT system designer. When it comes to businesses, a clear segregation is made among industrial IoT, consumer IoT, and smart agriculture in the IoT landscape, although the architecture remains very alike for the three. Before delving into the architecture, we would like to discuss a little about the aforementioned terms.

- **Industrial IoT (IIoT):** When IoT systems are designed in such a way that they optimize the processes or operations of factories, logistics, and other industrial use cases, it forms the scenario of IIoT. The hardware and software to be used and developed in these use cases should be of industrial standards and robust enough to sustain the load of the factory or assembly lines where they are deployed. For instance, MAN Truck and Bus, a German company, provides its customers with trackers which help in predictive maintenance by spotting potential failures and faults in the engine of the vehicle.
- **Consumer IoT:** This encompasses the IoT solutions developed for the everyday consumers at their homes, workplaces, etc. The term "smart devices" is most apt to be used here, as smart devices and consumer IoT systems are a parlance which can be interchangeably used in most cases. Various instances, such as smart fridges, smart homes, smart air-conditioners, and smart televisions, all fall in the domain of consumer IoT.
- **Smart agriculture:** IoT systems which help in improving the output and/or utility of the resources in farming and agriculture form this cluster of smart agriculture. For example, the use of smart irrigation is an end-to-end system which has sensors to understand the soil moisture levels and temperature and accordingly actuate by turning on and off the motor or the pump which irrigates the crops. This IoT system is not just increasing the yield of the crops but also enhancing the judicious use of scarce resources such as water. Another use case which we find to be very pertinent for the agrarian scenarios in emerging countries such as India is the deployment of a holistic smart system which senses the soil moisture, pH level, and light intensity from the farms spread across a district or a zone and then sends this data to the cloud in real time; this data can be accessed by the concerned authorities by virtue of web applications, thus helping them monitor the dire agricultural scenario in real time and thus allocate resources as needed by the farmers. Lots of start-ups are emerging in this area, which are innovating to solve the problems of both supply and demand of agriculture.

Coming back to our primary topic of discussion, let us now explore the IoT architecture from the dimension of an enterprise as shown in Figure 6.4. We can also witness the offerings made by enterprise-level IoT in Figure 6.5.

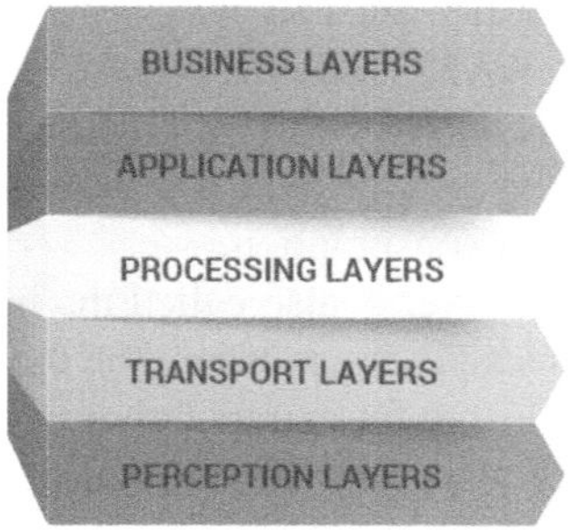

FIGURE 6.4 IoT architecture from a business perspective. (Community by NASSCOM

Let's understand each layer and its functionality in the overall IoT architecture for enterprises.

1. **Perception layers:** This is the layer which is the same as the physical layer of the developer's architecture. The sensing of data via sensors from the physical environment happens here. The use of various machine-to-machine protocols is also possible in this layer, wherein one smart device communicates directly with another smart device.
2. **Transport layers:** This layer is also known as the network layer; just like the gateway layer of developers' architecture, this layer is responsible for first receiving the data from the perception layer and then sending it to the upper layer through the use of internet and vice versa. This transportation is mainly done by networks, such as 4G/LTE (long-term evolution),

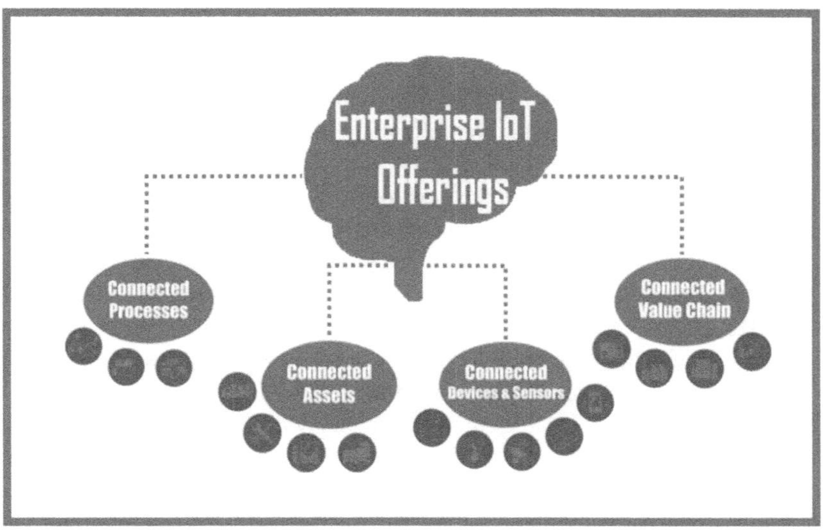

FIGURE 6.5 Offerings provided by Enterprise IoT.

is mainly done by networks, such as 4G/LTE (long-term evolution), LAN (local-area network), WAN (wide-area network), and SDN (software-defined network). 5G technology will become a game changer in this layer as it will offer higher speeds of data communication and also can offer devices with lower power consumption.

3. **Processing layers:** This layer is also called the middleware layer; it comprises the cloud systems and servers where the data is received via internet and then is processed further for extrapolating information out of the data. Data storage as well as data analytics are all done in this very layer. In the case of some decision logic which leads to actuation, this layer further sends the direction to the lower layers so that the actuators in the perception layer can act accordingly.
4. **Application layers:** This is the layer wherein the services offered by IoT interact directly with the end-users. The services related to application are myriad ranging from smart homes to individual smart appliances and smart devices.
5. **Business layers:** This layer is most pertinent to the enterprise which develops and/or deploys the IoT systems and services. The business models, profit plans, use case, and application all come under the purview of this layer. This is where we decide how we monetize the data so generated from the IoT devices. If some process is being optimized by some IoT system or some solution to a problem is offered, then the efficiency of the system is evaluated here.

6.5.2 IoT Architecture from the Perspective of Computation Capacity

Now, that we have understood the IoT architecture from the perspective of development as well as of an enterprise, let's understand the same from the angle of computation capacity or the processing power and capability of any microprocessor- or microcontroller-based system. Basically, the IoT systems reflect a very centralized structure, but the computation power and logic are distributed across various layers of the IoT architecture, shown in Figure 6.6. Broadly, the architecture comprises these three layers as mentioned in the following in terms of computation:

1. **Edge layer:** This comprises the edge devices or the end-user devices of the physical layer which are directly in contact with the real world; basically, they are billions in number. The memory and processing power of the edge devices are comparatively small, and hence they are restricted mostly to collecting the data and doing very minor computations and passing on the data to the upper layers for further processing. Basically, microcomputers, microcontrollers, mobile phones, and personal computers all are examples of edge devices.
2. **Fog layer:** This layer often merges directly with its upper layer of cloud computing, but in some use cases such as factories, which have multiple assembly lines, this layer comes in handy as one such fog computing device can be deployed per assembly line. This further helps in decentralizing the processing of the IoT system. Basically, these devices are computers with decent computation capability and are connected over a LAN to the edge

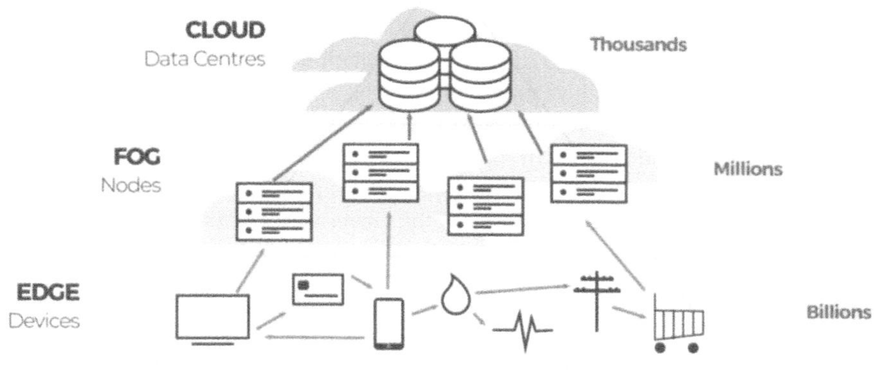

FIGURE 6.6 IoT architecture based on computation capacity. (Zigurat Innovation and Technology Business School[8].)

devices. They process the data to a degree of refinement before sending it to the upper layer of cloud computing.

3. **Cloud layer:** This layer consists of high-speed processors and servers located in a data centre which is maintained by an enterprise. The communication to the lower layers is only through the internet as the geographic degree of separation of the edge devices and fog devices from the cloud computers is very high. Here, high-end and faster processing computers are deployed, and the storage of tons and tons of data in degrees of petabytes is done. Data warehousing to data analytics and deploying user applications all occur in this layer. The popular examples are AWS (Amazon Web Services), Microsoft Azure, and others.

6.6 SMART CITIES

If one were to look back into the history of mankind, one would find that civilizations thrived along the river banks, and some of the oldest cities such as Varanasi, India, are still there. But today with the advancement of technology and the dire need for digital transformation, the landscape is not confined along the banks of rivers. With a global population of 7.9 billion, we have numerous cities spread across the world; some harbour very less people in them, while some are so densely populated that it leads to the scarcity of resources at various times of the year.

We all would agree that cities are the building blocks of a nation. If the availability of resources in cities is met, then eventually the needs of an entire nation's population are catered to. When we discuss IoT, one of the major lookouts is for device-to-human ratio, that is to say, the number of smart devices available per person. But these smart devices with individual consumers are not the complete picture in themselves. With this thought of completing the equation emerges the outlook of forming smart cities. A smart city is an urban area with proper infrastructure and resources necessary for better means of living of its citizens. A smart city employs various digital and electronic methods to manage assets, resources, and services in the best possible manner. In this chapter, we will showcase IoT as the fundamental backbone of smart cities and digital transformation as the primary pillar which lays the foundation of the smart cities. We will be discussing various matrices and approaches taken up by some private companies and governments across the world with regard to smart cities. We will focus on some of the important data and statistics when it comes to smart cities in the Indian context, as it has been taken up by the Ministry of Housing and Urban Affairs of the Government of India.

6.7 AMALGAMATION OF IoT IN SMART CITIES

Definitely, the smart city is not just about developing and deploying digital systems for the use of residents but also the core infrastructure elements, such as water supply, sanitation, transport, health, and education. In this section, we mainly focus on the phases and parts wherein IoT is the major transforming agent and the backbone of the smart city ecosystem. If we consider the Indian context, there are broadly two important pillars of any smart city; the first is the core infrastructure and the other is smart solutions.[2]

Core elements of infrastructure in a smart city:

1. Adequate and proper supply of water for all purposes.
2. Waste management system and good sanitation facilities.
3. Electricity should be available without interruptions.
4. Affordable and timely urban transport and solutions of mobility.
5. Sturdy digital infrastructure and ICT networks.
6. For the poor citizens, affordable housing facilities.
7. Participation of citizens in governance.
8. Environmental sustainability.
9. Healthcare and education facilities.
10. Safety and security of all the citizens.

Smart solutions component of a smart city:

1. Citizen services and e-governance:
    a. Electronic service delivery.
    b. Public information and grievance redressal.
    c. Smart fire hazard alert system for the fire department.
    d. Citizen engagement.
2. Water management:
    a. Monitoring of water quality.
    b. Effective pipeline leakage detection and preventive maintenance.
    c. Smart water supply meters.
3. Waste management:
    a. Conversion of wastes to energy, fuel, and compost.
    b. Effective recycling.
    c. Effective reuse of resources.
    d. Smart waste water treatment.

4. Energy management:
   a. Smart electricity meters.
   b. Smart grids.
   c. Electricity theft detection and prevention.
   d. Energy-efficient technologies and green buildings.
   e. Use of renewable energy sources.
5. Urban mobility:
   a. Intelligent traffic management.
   b. Accident detection and prevention systems.
   c. Smart parking systems.
   d. Smart tracker systems for public transports such as trains and buses.
6. Healthcare and education:
   a. Telemedicine.
   b. Tele-education and e-learning (ed-tech).
   c. Smart handling of medical emergencies.
7. Crime management and prevention:
   a. Video surveillance system and crime monitoring.
   b. Online availability of criminal records as per the geographic location.
   c. Quick response system for women's safety.

When it comes to the smart solutions component of a smart city, most of the solutions offered are IoT systems and their derivatives. Thus, smart cities can also be viewed as a cluster of different IoT systems ranging from consumer IoT to large-scale industrial IoT. Any smart city should ensure two things: first, the proper supply of power or electricity through the city, and second, proper network connectivity to the internet or a WAN across all areas of the urban and semi-urban landscape. These two factors are an unsaid imperative towards implementing IoT-based systems and solutions throughout.

We will be discussing different use cases which are already implemented in various smart cities around the world as well as some use cases which are yet to be made feasible for use but can be implemented in future. Various solutions, such as smart water supply meters, smart electricity meters, smart grids, smart urban transport systems, smart parking systems, and intelligent traffic management systems are all fundamentally IoT-based systems.

ICT of the smart city has to work in tandem with the connected devices so as to form a complete ecosystem of smart devices. All the layers of IoT architecture are actively involved as the building blocks of a smart city. Thus, the threading together of smart cities, smart factories, smart hospitals, and smart highways is possible by virtue of IoT. The overall economy of a city can be boosted by implementing a proper IoT ecosystem there as captured in Figure 6.7.

With so many connected devices, there will be high volumes of data generated that can be stored in a data centre which can also be part of the smart city; the data so collected from various smart devices can be analysed further so as to help in improving the quality of life in a smart city and also ensuring better governance. Data from avenues such as healthcare, criminology, transport, food consumption, and water usage become very easy to gather, thus ensuring faster statistical and data

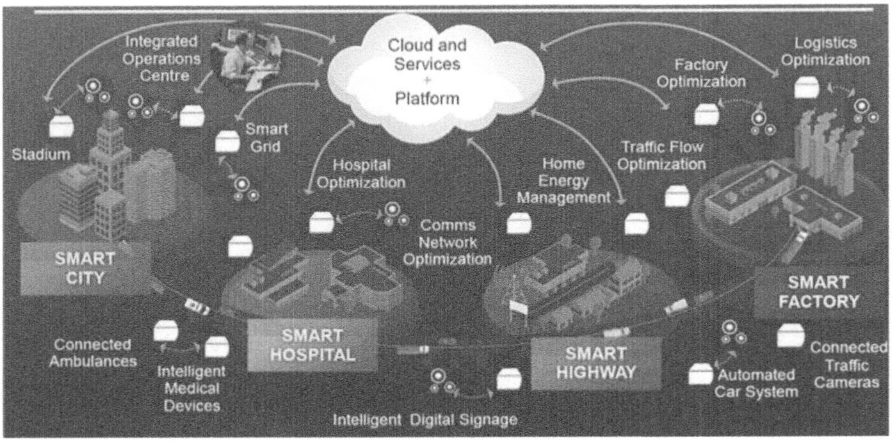

FIGURE 6.7 An entire smart ecosystem created by virtue of IoT and connected devices. (Profit from IoT.com, iot.electronicsforu.com.[3])

science-related approaches towards solving the social problems faced in a particular geography. The assets and resources become easy to track and manage for a city.

6.8 CHALLENGES IN IMPLEMENTING IoT FOR SMART CITIES

There are a good number of roadblocks and hurdles when it comes to the deployment of IoT systems in smart cities; here we are going to discuss the major challenges faced by any organization when providing IoT-based solutions for smart cities.

1. **Lack of high-speed internet connectivity:** In many countries, especially the emerging economies, not all the cities have proper communication infrastructure in place. In fact the latest communication technologies such as 4G/LTE were rolled out in countries like India at a much later stage as compared to the developed countries which had access to it four years prior. Even though India has managed to provide internet access to 77 crore users out of a population of 138 crores, robust internet connection to every nook and corner of the country remains a distant dream. This causes delay in relaying the information and thus becomes a bottleneck in the implementation of an ecosystem of connected devices. For a smart city, network downtime will lead to serious losses, both monetary and operational.
2. **Poor rates of digital literacy:** The connected devices will require a user who is also digitally literate. Most smart devices demand a smart user for them given certain complexities in operating these devices. Given a developing economy like India where only 38% of households are digitally literate, it becomes an impediment to achieve fullest use of the connected resources, which reflects badly on the cost to benefit analysis of the IoT systems. Even among the urban households, only 61% are digitally literate in India.

3. **High cost of capital:** The initial costs of establishing the infrastructure such as cellular network towers and optical fibre networks for internet connectivity need huge capital, and high expenses are to be borne, which is only possible if the government or some large private enterprise funds the projects. Again, speaking from an Indian perspective, the tier 2 and tier 3 cities have to fall out in the development of IoT systems because of this. IoT-based start-ups too have to indirectly depend on the telecom goliaths to test their devices in real time.
4. **Time to monetize from the IoT systems:** In general, the easiest way to monetize or reap profits from an IoT system is through the route of data mining. Usually, the time taken to monetize from the connected ecosystem is long, perhaps more than two years in most cases, and thus short-term profits have to be sacrificed. Thus, for smaller players, howsoever innovative they may be in terms of developing and designing IoT solutions, undercapitalization remains a lurking problem.
5. **Cybersecurity and other vulnerabilities:** There's a very famous saying in computer networking: 'only the computer system which is not connected to any network is the safest'. Yes, any and every system on any network is exposed to various attacks from hackers and exploiters of vulnerability, which would be not good if some unauthorized group or individual gets access to any data to which they are not entitled to; thus, alongside IoT developers and engineers come the added expense of cybersecurity experts and cybernetics professionals.

6.9 FRAMEWORK AND ASSESSMENT MATRICES OF SMART CITIES

There are various frameworks available which help in assessing and developing plans for a smart city; in these same lines, different matrices are also available for evaluating the performance of a smart city. Here, we will discuss the frameworks and matrices developed by two large international companies.

1. McKinsey Global Institute: Smart Cities[4]
2. Siemens Atlas of Digitalization[5]

We will delve into the earlier frameworks and their related layers and assessments in detail alongside showcasing some of the data which the companies have used for various cities of the world.

1. McKinsey Global Institute: Smart Cities – Digital Solutions for a more liveable future

First, we will understand what the layers are as shown in Figure 6.8, which help in building a smart city or simply answer the question of what does a smart city comprise. Hence, this will form a framework for a smart city. Then we will delve into the various aspects of urban quality of life which are improved by the advent of smart cities and connected devices. Thus, it will be an assessment matrix for the impact of the smart ecosystem.

FIGURE 6.8 The three layers which are imposed on a city to become a smart city. (McKinsey Global Institute analysis.)

- The traditional physical as well as social infrastructure of a city forms the layer zero of the smart city. The other layers are now super-imposed on it like a lasagna so as to turn a regular city into a smart one. The core infrastructure elements, such as electricity supply and water supply, which have been discussed earlier in the chapter, become an integral part of this layer. The better the availability and quality of the core infrastructure elements, the more robust and sturdier this layer becomes.
- The first layer, as is conspicuous from Figure 6.8, is the Tech base, which is where all the smart devices and their related communication networks come into picture. The various elements of the first three layers of IoT systems, the physical layer, gateway layer, and cloud layer, are the ones which form this layer. Sensing devices, servers, and networking devices all are part of this.
- The second layer corresponds to the Application layer of IoT architecture. Here various use cases of smart devices and their related data, which has been processed to produce the required output, are present. Basically, the human and IoT system interaction and the impact of the use case are covered here. The interaction can be direct to a consumer, such as smart electricity metering, or can be indirect such as smart grid.
- The third layer is the Adoption layer, which actually tells us the degree to which the citizens of a smart city have adopted a particular use case which has been implemented. This is evaluated by the frequency of use of the given IoT system by the end-users and also eventually if it leads to some decision or behaviour changes of the user, then it is said to have made a deeper impact. For instance, if by use of smart-fridge people get used to checking their list of food items on the mobile app every time they visit a store, it shows that they have highly adopted.

Now, that we have understood the layers suggested by MGI, let's proceed to the assessment matrix they have employed. For evaluating the outcome of deploying IoT systems in smart cities, MGI reflects on seven different quality of life aspects as

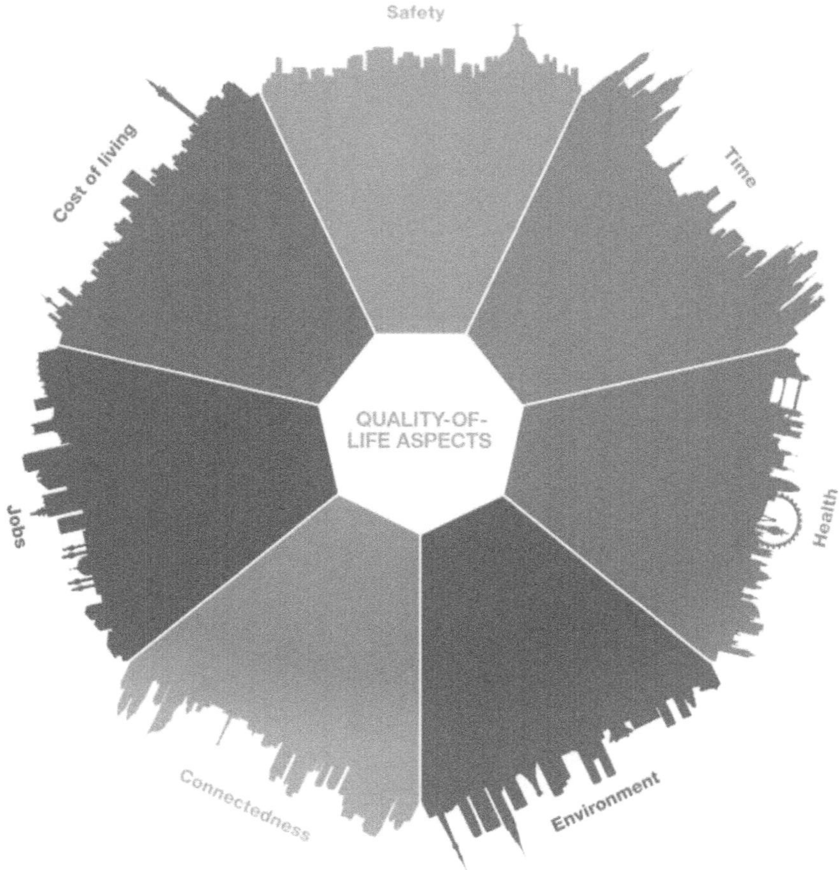

FIGURE 6.9 MGI's seven dimensions of quality of life impacted by the smart city. (McKinsey Global Institute analysis.)

shown in Figure 6.9. Smart solutions are offered for each aspect, and their result is noted.

Although the impact on these seven dimensions of quality of life depends on the legacy infrastructure which is already available in a given city, we will showcase certain smart solutions for each aspect and also see the degree of impact these solutions can possibly have for any given smart city in Table 6.1.

2. Siemens – The Atlas of Digitalization

Based on the sub-themes of Expo 2020 Dubai (held in the year 2021 due to the pandemic in the year 2020), the German technology and innovation giant Siemens came up with its own framework to realize the digitalization readiness of various smart cities of the world, as well as a matrix with the same parameters as the framework to gauge the potential of the global smart cities.

TABLE 6.1
Impact of some smart solutions on the seven quality of life aspects

Dimension	Smart Solutions	Possible Impact
Safety	• Smart traffic management • Data-driven policing (crime-monitoring and surveillance) • Optimized emergency response	• Reduction in criminal activities by 30–40% • Saving of 30–300 lives per year • Faster arrival of first respondents by 2–17 minutes
Time and convenience	• Apps for public transport • Smart traffic monitoring and management • Newer, advanced options of mobility	• The average commutation becomes 15–20% shorter • 15–30 minutes/day are regained by workers • Productivity improves as gridlock is reduced
Health	• Better treatment of chronic diseases • Public health interventions as driven by the data • Telemedicine and enhanced patient experience by digital tools	• Disease burden reduces by 8–15% • 8–15% decrease in ill effects of air pollution • Good health and wellness are beaconed from cities
Environment	• Monitoring systems for air quality • Optimizing use of energy • Tracking system for essential resources such as electricity, water	• 10–15% reduction in emission of greenhouse gases • Reduction in annual waste per person by 30–130 kg • Daily 25–80 L of water saved per person
Social connectedness and civic participation	• Social media apps • Grievance redressal apps and apps to connect individuals with government • Neighbourhood connectivity apps	• Real-world connections translated from digital world • People can voice their opinions in governance and planning • Neighbourhoods linked virtually
Jobs	• Online hiring platforms and digital career centres • Small businesses can have digital administrative process • Ed-tech	• Employment rates up by 1–3% • Small business and start-up friendly ecosystem • Upskilling of the citizens
Cost of living	• Digitized land-use and permitting processes • Dynamic electricity pricing and digital usage tracking • Electronic management and tracking of resources such as water	• 1–3% reduction in citizen expenditures • Affordable housing • Increased supply and availability of resources

Source: McKinsey Global Institute.

As per Siemens, a smart city is one which understands how to use technology to 'deliver Sustainability, improve Mobility, and create Opportunity'. Let's understand the three pillars of the framework used by Siemens:

1. **Sustainability:** In this case, sustainability basically refers to the green approaches taken up by a city; that is, how ready is a smart city to work towards the conservation of environment, how well they are combatting climate change, and what are the various approaches and steps taken up in that direction. The way a city handles its resources, such as air quality, water, and energy, all add up to sustainability goals. Even the development of green spaces (parks, tree plantings, etc.), decarbonization and use of renewable sources of energy, and the use of efficient and smart lighting are part of this domain.
2. **Mobility:** Mobility answers the question of how easy it for the citizens of a smart city is to move around. As the population of a smart city grows, it should be ready to provide good means of urban transportation to the people. The better the transportation infrastructure of a city, the better its mobility. A smart city is supposed to undertake rapid transit means of transport, such as monorails, and fast lanes for cars. Various IoT solutions, such as intelligent traffic management systems, all enhance the mobility of a smart city. Even the choice of the residents to use electric vehicles over combustion engines and reduce emissions, the increased usage of public transport, shared bicycles from various apps, shared cars, and other practices all deliver to the mobility paradigm. Various solutions such as centralized tracking and monitoring of public transport and its display to commuters via mobile apps are add-ons to this pillar.
3. **Opportunity:** The third pillar of opportunity is directly related to the economy of a smart city. As per Siemens, every smart city should be able to provide some competitive advantage to individuals and companies for being there, rather than being located in some other location. The survival of businesses in a local economy, the availability of different career opportunities, and the continued availability of a diverse talent pool help in increasing the stakes of opportunity for a smart city. If an ecosystem of innovation is provided by a smart city, it would definitely lead to the emergence of various start-ups in its vicinity. Investments made in the frontiers, such as free trade zones, start-up incubators and accelerators, tech hubs, and universities, all help in attracting both new individuals and companies to the smart city. Digital inclusion of the citizens and digital transformation of traditional sectors such as mining also add to opportunities.

To visualize these three pillars of a smart city in a better manner, Siemens has come up with its own visualization tool called Ciuty as shown in Figure 6.10, which is hosted on their website for smart cities: Atlas of Digitalization.

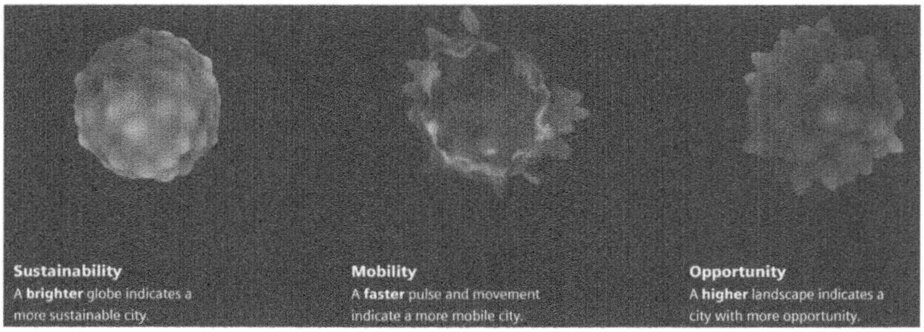

FIGURE 6.10 Ciuty, a 3D visualization tool of Siemens to convey the three pillars of a smart city. (Siemens Atlas of Digitalization.)

Siemens Atlas of Digitalization assessment matrix for smart cities is based on the three pillars:

1. **Digitalization readiness:** This is a score on a scale of 0–10, where 10 conveys the maximum. This tells us how well equipped a city is to become a smart city and how well it can embrace the use cases of connected devices and a smart city ecosystem. It takes into consideration the talents and merits of the citizens as well as the existing infrastructure conditions of the city.
2. **Digitalization potential:** It is also rated on a scale of 0–10, wherein 0 denotes the minimum and 10 is the maximum. This parameter encompasses the prospective and the scope of a smart city and its ability to transform all its resources and assets as smartly as possible in the future. It definitely encompasses the current smart city and IoT implementation projects taken up, and the possible degree of impact digital transformation will have in the three pillars of the smart city.

In Table 6.2, we can see the scores given by Siemens to some of the major smart cities of the world, on the aspects of digitalization readiness and digitalization potential.

One interesting thing to mention here is that Digital Readiness mostly encompasses the past and current infrastructure and connected devices ecosystem of a smart city, whereas the Digital Potential mainly focuses on the future prospects and the adaptability of the smart city towards digital transformation.

Apart from the two frameworks and assessment matrices which we have discussed in this chapter, there are also other frameworks for smart cities. As the smart city is an evolving space, the researchers and various organizations are coming up with newer frameworks and matrices for assessment. Here, we would like to mention some other prevalent frameworks as well for smart cities:

1. **Technology framework:** As a smart city heavily depends on the deployment of technology, especially IoT and other ICT, frameworks are designed from the lens of the tech infrastructure,[6] which are to be placed in the smart city ecosystem.

TABLE 6.2
Digital Readiness and Potential Scores of Various Smart Cities across the Globe by Siemens

Smart City (Chosen by Siemens)	Digitalization Readiness (Score Out of 10)	Digitalization Potential (Score Out of 10)
Bengaluru	4	5
Dubai	6	3
Johannesburg	2	5
Cairo	3	5
Istanbul	5	4
Berlin	6	2
Munich	5	2
London	8	3
Sao Paulo	5	4
Buenos Aires	3	4
New York	6	3
Toronto	7	3
Los Angeles	7	5
Melbourne	6	3
Taipei	6	2
Hong Kong	6	4
Singapore	8	2
Bangkok	4	3

2. **Human framework:** The ultimate impact of smart cities is on its users, that is, the citizens and visitors to the city. Thus, various aspects, such as economy, human support systems, and knowledge networks, become indicators of the successful impact of smart cities.[7] The dimensions of creativity, learning, humanity, and knowledge should be captured in any smart city.
3. **Energy framework:** Various smart city solutions, such as smart lighting of the city, intelligent traffic management systems, and other such large IoT systems, require a huge amount of power to function; thus, large electricity and power distribution corporations become instrumental in providing a blueprint for the execution of such projects which are to be implemented city wide.
4. **Data management framework:** With the advent of smart cities, there will be high volumes of data to be processed as well as stored, and the level of data varies from one smart device to another; thus, the perspective of data management is also another perspective.

6.10 SMART CITY FROM AN INDIAN CONTEXT[9]

On 25 June 2015, Smart Cities Mission was launched by the Government of India, with the objective of promoting and developing inclusive and sustainable cities that provide a decent quality of life to their citizens and have a core infrastructure, a sustainable and clean environment and witness the application of 'Smart Solutions'.

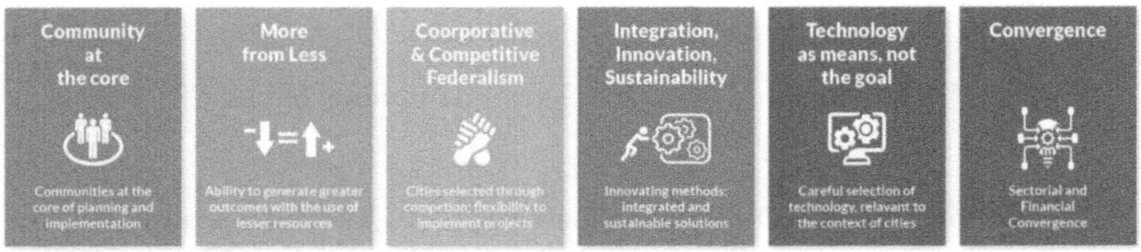

FIGURE 6.11 The six principles for smart city development and planning by the Government of India.

After a two-stage competition, 100 cities across India have been selected for them to be developed as smart cities. Operated as a scheme which centrally sponsored, financial support up to the amount of Rs. 48,000 crores will be provided by the Union Government of India over a period of five years, hence on an average annually Rs. 100 crores per city. The Government of India has not laid out any particular template for the smart cities. However, the local as well as the central government try to align them with these six principles as shown in Figure 6.11, when it comes to planning and development of the Indian smart cities:

1. Community at the core of planning and developing smart city solutions.
2. Generating more by consuming less of resources.
3. Cooperative but also competitive federalism, to promote innovation and knowledge sharing between the competing cities.
4. Integration of all the solutions towards achieving a larger goal for the smart city as a whole while promoting innovation and sustainability.
5. Technology should be the means of achieving the desired goals and digital transformation should not become the goal in itself.
6. Convergence of the solutions and infrastructures towards uplifting economy and quality of life.

The development of smart cities is one of the top priorities for Indian government, as the expenditure on smart cities has exceeded INR 58,500 crores (as shown in Figure 6.12).

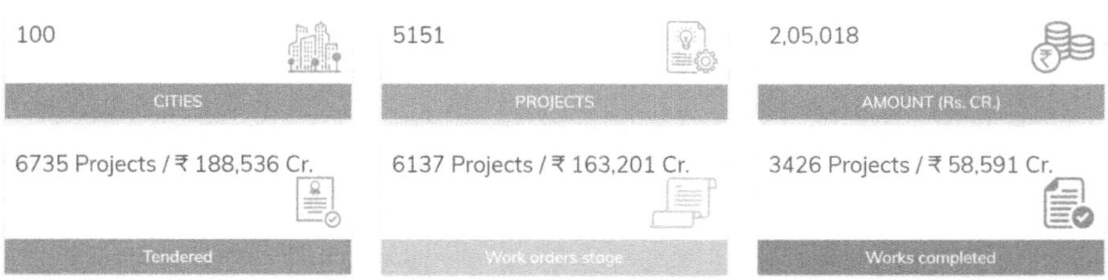

FIGURE 6.12 Total expenditures on Smart City Mission of Government of India (as on 27 February 2022). (Smart City Mission Transform Nation, Dashboard, MoHUA, Government of India.)

6.11 IMPLEMENTATION OF SMART CITY MISSION IN INDIA[9]

The implementation of this at the city level is done by virtue of an SPV (special purpose vehicle) created for this purpose. The Government of India has financed huge sums so far for the SPVs as can be witnessed from Figure 6.12. It is the major driving body for the execution of the plans and development activities.

As per the Government of India, the following are the key functions and responsibilities of SPV:

1. Project sanctioning and approvals, including the appraisals that are technical in nature.
2. Operational discretion for the execution of the Smart City Proposal (SCP).
3. Ensures that steps are taken to adhere to the requirements of Ministry of Urban Development when comes to the execution of the Smart Cities programme.
4. Mobilization of resources with adherence to timelines.
5. If any third-party Review and/or Monitoring Agency furnishes a report, SPV has to take action on it and consider its approval.
6. Supervision of activities related to building capacity.
7. Linking together of academia and other organizations and reaping benefit out of it.
8. Make sure that the projects get completed as per the decided timelines.
9. Review various undertakings of the Mission inclusive of the finances, execution of project activities, and drafting of SCP and coordinating activities of various ministries and other missions/schemes.
10. Reviewing and supervision of various matters related to quality control and taking action accordingly.
11. Create Public Private Partnership, subsidiaries and get into joint ventures wherever possible for better implementation of the projects.
12. Determination as well as collection of charges, taxes, etc. as permitted by the ULB (Urban Local Body).

6.11.1 Structure of SPV

It is a Limited Company as per the Companies Act, 2013. The city level SPV is jointly promoted by the State/Union Territory and the ULB, with a 50:50 equity shareholding between them. This nature of shareholding has to be maintained throughout. Given the terms that State/Union Territory and ULB together have majority shares of the SPV and have its control, and their shareholding is equal to that of each other, then private institutions or financial bodies can be considered to obtain equity stake in the so formed SPV. Apart from the equity, the State/Union Territory can provide grants to the SPV to ensure there is no shortage of funds, thus fulfilling the State Government's responsibility in the Smart Cities Mission. The SPV is headed by a CEO, and its board has nominees from the ULB, State Government, and Central Government.

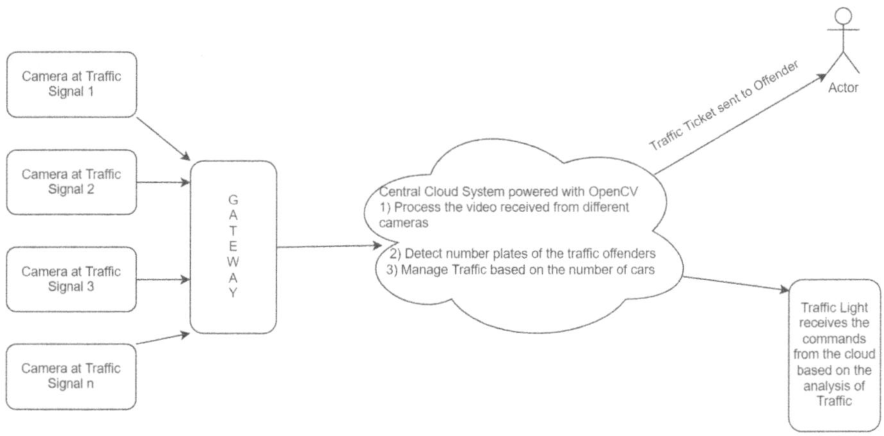

FIGURE 6.13 Use-case diagram of smart traffic monitoring and management system.

6.12 IoT FOR SMART CITIES COMMON USE CASES

This section highlights the use cases of smart cities. First is the smart traffic monitoring (as shown in Figure 6.13), second is smart energy meter (as shown in Figure 6.14), and third is smart water supply meter (as shown in Figure 6.15). Smart traffic monitoring uses the advanced technologies and data analysis techniques to monitor and manage traffic on roadways more efficiently and effectively. It involves the collection, processing, and analysis of real-time traffic data to improve traffic flow, enhance safety, reduce congestion, and provide valuable information to both transportation authorities and commuters. Smart traffic monitoring plays a crucial role in improving transportation systems in urban areas, making them more efficient, safe, and environment-friendly. It also contributes to the overall quality of life for residents by reducing commute times and stress associated with traffic congestion.

A smart energy meter measures and records energy consumption in a more advanced and intelligent way compared to the traditional analog energy meters. These meters are a key component of modernizing and improving the efficiency

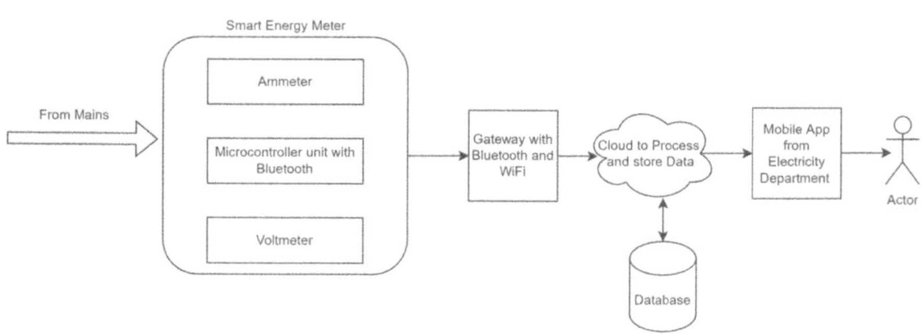

FIGURE 6.14 Use-case diagram of a smart energy meter for electricity billing and monitoring.

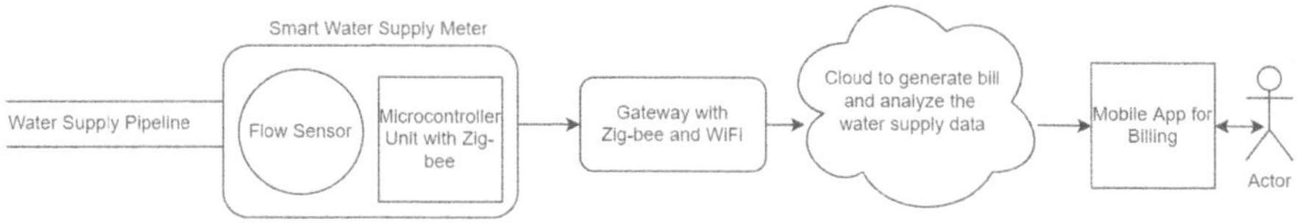

FIGURE 6.15 Use-case diagram of a smart water supply meter for billing and monitoring of water.

of energy distribution and consumption. Smart energy meters are a crucial component of modernizing the energy infrastructure and enabling the transition to a more efficient and sustainable energy system. They empower both utility companies and consumers to make informed decisions, optimize energy use, and reduce energy waste. Additionally, they play a vital role in supporting the integration of renewable energy sources into the grid.

Finally, the smart water supply meter for water, is used to measure and manage water consumption more efficiently compared to traditional water meters. These meters leverage technology to collect and transmit water usage data, providing numerous benefits to both utility companies and consumers. Smart water supply meters are a critical component of modernizing water distribution systems, improving water management, and promoting water conservation. They help utility companies optimize their operations, enhance customer service, and reduce water losses, ultimately leading to more efficient and sustainable water supply systems.

1. Smart traffic monitoring and management system.
2. Smart energy meter.
3. Smart water supply meter.

6.13 CLOSING NOTES

In this chapter, we discussed what IoT is and its related architecture and saw how it acts as a backbone for the design and implementation of the smart cities. Going ahead, every designer and planner of smart cities has to bear in mind the multifaceted role of IoT-based smart solutions in the effectiveness of a smart city. This chapter will help researchers who wish to dig further into smart cities and related use cases. The discussions in this chapter give an immense conceptual understanding of IoT and its architecture from different perspectives. Students, professors, and researchers who wish to understand the role IoT plays in a smart city will benefit from this chapter. This chapter acts as a precursor and explainer for anyone who wishes to embark on a journey to understand smart cities in a better manner. People who do not have the background knowledge of IoT and smart cities can use this as the starting point to dive deeper into these domains. Smart city planners, government agencies, and various corporate bodies, who are involved in the design and development of smart cities, will most certainly find the frameworks and assessment matrices very handy tools for their body of work. Agencies who wish to analyse the effectiveness and impact of smart cities can easily use the points discussed in this chapter. The use cases discussed are equally relevant for engineers and business solution providers who wish to contribute to the field of connected devices and smart cities.

There are things which are beyond the scope of this chapter and hence were not captured. The granularity of engineering challenges of IoT and other solutions such as blockchain technology for smart cities are not discussed here. IoTs and the concept of smart cities are all in their nascent stage of development; hence, we cannot provide a one-size-fits-all solution for implementing IoT in smart cities. Other various fronts of smart cities such as sustainability are not covered exhaustively in this chapter. This book lays down a framework and a good starting point for those who desire to contribute to deploying IoT-based solutions to increase the effectiveness of smart cities in improving lives.

In the coming days, the detailed study of individual use cases for smart cities can be taken up as avenues for publishing papers and developing patents. Business dimensions, from modelling to monetization of smart solutions, are something which will be very useful for both industry and academia and can be taken up as a successor to this chapter. As new solutions emerge and newer smart cities develop, more data will be available to fit into the framework and assessment matrices of this chapter. The real-world smart solutions offered by virtue of this chapter can easily be translated into case studies for further enhancing the robustness and resilience of the frameworks. It will also help in understanding the challenges encountered in large-scale IoT deployment in a more refined and finer way.

6.14 KEYWORDS AND DEFINITIONS

IoT: IoT, the technology and process of connecting things which are usually not connected to any computer network to the internet.

Microcontroller: A lower capacity and smaller computation device which has the ability to interact with the environment by virtue of inputs and outputs in the form of electronic signals.

Sensor: An electronic device which can sense a particular physical phenomenon from the real world and convert it into electronic signals.

Actuator: An electronic device which takes inputs in the form of electronic signal and then produces output in the form of some physical phenomenon.

Cloud computer: Computer servers which are geographically located in a different location but can be accessed by virtue of internet so as to process data and/or store data from other systems; usually they have very high data-processing capability.

Edge device: Computing devices with lower computation and data-processing capability, usually used to sense and collect data.

Fog device: Middle-level computer systems, which have moderate-to-high data-processing power; usually a cluster of these form a cloud.

Smart device: An electronic device with communication and connectivity capabilities and ability to process and store small amounts of data.

Gateway: A networking and connectivity device which helps in transferring data from one device in a smaller network to devices connected to a larger network area.

Microcomputer: A small computation device with very low storage and data-processing capabilities.

Digital transformation: The process of bringing change or optimizing a system or a process by an adoption of technology and use of digital platforms.

Digitalization: The process of bringing change in the existing ways of social life by direct and indirect exposures to digital technology.

Digitization: The process of converting anything, be it some data or information, into a digital format.

Smart city: An urban area which has good core infrastructure alongside large-scale digital adoption and connected devices to improve the quality of life of its citizens.

REFERENCES

An Introduction to IoT Architecture and Its Benefits for Enterprises!. (2022). Retrieved 31 January 2022, from https://community.nasscom.in/communities/emerging-tech/iot-ai/an-introduction-to-iot-architecture-and-its-benefits-for-enterprises.html#:~:text=In%20essence%2C%20an%20IoT%20architecture,system%20through%20protocols%20and%20gateways.

Ash, B., & Chandrasekaran, S. What the "Smart" in Smart Cities Is All About. 2017. Available online: https://beyondstandards.ieee.org/smart-cities/smart-smart-cities/

Batty, M., Axhausen, K. W., Giannotti, F., Pozdnoukhov, A., Bazzani, A., Wachowicz, M., ... & Portugali, Y. (2012). Smart cities of the future. *The European Physical Journal Special Topics*, *214*(1), 481–518.

Caragliu, A., Del Bo, C., & Nijkamp, P. (2011). Smart cities in Europe. *Journal of Urban Technology*, *18*(2), 65–82.

Chatterjee, S., Kar, A. K., & Gupta, M. P. (2018). Success of IoT in smart cities of India: An empirical analysis. *Government Information Quarterly*, *35*(3), 349–361.

Chourabi, H., Nam, T., Walker, S., Gil-Garcia, J. R., Mellouli, S., Nahon, K., ... & Scholl, H. J. (2012, January). Understanding smart cities: An integrative framework. In *2012 45th Hawaii international conference on system sciences* (pp. 2289–2297). IEEE.

Ismagilova, E., Hughes, L., Dwivedi, Y. K., & Raman, K. R. (2019). Smart cities: Advances in research—An information systems perspective. *International Journal of Information Management*, *47*, 88–100.

Jin, J., Gubbi, J., Marusic, S., & Palaniswami, M. (2014). An information framework for creating a smart city through internet of things. *IEEE Internet of Things Journal*, *1*(2), 112–121.

Khan, A., Aslam, S., Aurangzeb, K., Alhussein, M., & Javaid, N. (2022). Multiscale modeling in smart cities: A survey on applications, current trends, and challenges. *Sustainable Cities and Society*, *78*, 103517.

Khan, M. A., & Salah, K. (2018). IoT security: Review, blockchain solutions, and open challenges. *Future Generation Computer Systems*, *82*, 395–411.

Lea, R., & Blackstock, M. (2014, December). City hub: A cloud-based IoT platform for smart cities. In *2014 IEEE 6th international conference on cloud computing technology and science* (pp. 799–804). IEEE.

Letaifa, S. B. (2015). How to strategize smart cities: Revealing the SMART model. *Journal of Business Research*, 68(7), 1414–1419.

Rajab, H., & Cinkelr, T. (2018, June). IoT based smart cities. In *2018 international symposium on networks, computers and communications (ISNCC)* (pp. 1–4). IEEE.

Samuel, S. S. I. (2016, March). A review of connectivity challenges in IoT-smart home. In *2016 3rd MEC International conference on big data and smart city (ICBDSC)* (pp. 1–4). IEEE.

Saravanan, K., Julie, E. G., & Robinson, Y. H. (2019). Smart cities & IoT: evolution of applications, architectures & technologies, present scenarios & future dream. In *Internet of things and big data analytics for smart generation* (pp. 135–151). Springer, Cham.

Sterbenz, J. P. (2017, September). Smart city and IoT resilience, survivability, and disruption tolerance: Challenges, modelling, and a survey of research opportunities. In *2017 9th International Workshop on Resilient Networks Design and Modeling (RNDM)* (pp. 1–6). IEEE. Alghero, Italy, 4 September 2017; pp. 1–6.

Zanella, A., Bui, N., Castellani, A., Vangelista, L., & Zorzi, M. (2014). Internet of things for smart cities. *IEEE Internet of Things journal*, *1*(1), 22–32.

7 Building AI-Enabled Smart Cities for Children with Autism Spectrum Disorder

A Parent's Perspective

Debarati Basak[1], *Vasundhara Kaul*[1], *Suman Ananthanarayan*[2], *Arshia Kaul*[1], *and Nikita Kalloorthottiyil Chacko (RECE)*[3]
[1]Carpediem EdPsych Consultancy LLP, Bengaluru, Karnataka, India
[2]Sree Narayana Guru College of Education, University of Mumbai, Mumbai, Maharashtra, India
[3]RECE, Dufferin Corporate Centre, London, Ontario, Canada

7.1 INTRODUCTION

Technology has become a critical part of life. With the all-pervasive nature of technology, many things have changed: the way we work, communicate and behave with each other. There have been many changes in all different areas. The constant changes in technology have transformed how the world functions (Russell and Norvig, 2003). Incorporating the use of technology in day-to-day activities is going to make a significant difference to all. The vision and mission of many countries are to develop smart cities for the development and growth of the country. Many researchers have discussed the concept of smart cities and the extensive use of technology in these smart cities for overall growth and development. Meijer and Bolívar (2016) defined the smart city from the perspectives of technology, human resources and collaboration. Adding to this is a fourth holistic approach which emphasises the need for the interactions between the first three perspectives.

Smart city scientists and practitioners follow two main perspectives. First is the academic literature which attempts to transverse beyond sector-specific approaches and proposes a more comprehensive conceptualisation of the smart city. The second approach has developed in the practical world through sector-specific initiatives (Fernández-Güell et al., 2016; Mattoni et al., 2015). Most implementation of smart cities remains restricted to these sector-specific projects given the limitations of governance and financing tools. It becomes essential to bridge the gap between

DOI: 10.1201/9781003225317-7

the comprehensive and the theoretical perspective and the sector-wide implementation of the smart city concept. Stakeholder involvement and engagement in decision-making is essential for smart governance and the critical element for becoming a smart city (Albino et al., 2015; Nam & Pardo, 2011a,b). However, stakeholders reveal different visions of the smart city in their discourses (Fernandez-Anez, 2016). There are also differences between the image of the smart city and its implementation (De Santis et al., 2014) and between the vision of the stakeholders in smart city development and the initiatives carried out (AlAwadhi & Scholl, 2013).

de Oliveira Neto (2018) has introduced the concept of "Inclusive Smart City (ISC)" to support the access and use of urban technologies by all citizens (including people with disabilities and seniors/older adults). He identified two requirements that any ISC initiative must have and satisfy as follows: governance and technological infrastructure. Governance requirements are visions and objectives concerning demands from underrepresented populations, whereas technological infrastructure requirements denote characteristics that can facilitate their accessibility, protect privacy and ensure user-friendly experiences.

Disability in India is a complex issue; it overlaps with other difficult challenges such as poverty, low employment rates, low literacy and widespread social stigma. Children with disabilities are five times more likely to be out of school than the average child (Fernández-Güell et al., 2016). Therefore, it would make the development of ISC in India even more complex. Innovations in science and technology like assistive devices and tools can go a long way to help persons with special needs. These can help children with special needs acquire the skills, education and training to become employable and empowered. The joy of an independent, empowered and fulfilled life should be made accessible to all those who are willing to work. The question to ask is what might the steps be to develop such an ISC in India.

When looking into the topics of research priority for families of individuals with autism spectrum disorder (ASD), similar trends have been observed across the globe. It has been noted that overall health and well-being; inclusive education and support from teachers and staff; recognising and supporting anxiety usually caused as a result of noise, crowding and overstimulation; and community awareness (Clark & Adams, 2020) are some of the top priorities. Besides these, in countries like Canada, financial aids and long-term housing situation for individuals on the spectrum have been a major concern for parents.[1]

Artificial intelligence (AI) can make the goal of creating an ISC possible. When we consider AI approaches and students with different abilities and needs, one can observe various new trends. Multiple tools are now available to help improve children's lives in educational or professional environments and the home setting. Researchers have also wanted to understand the different human intelligence forms. In the quest to understand human intelligence, AI was developed. AI is defined as the design and study of intelligent agents and ensures success rates (Russell and Norvig, 2003). One of the practical uses of AI is in education, and there is also successful implementation of AI.

Moreover, as there is an emphasis on the inclusion of all types of learners, AI plays an important role in catering to the varied needs of diverse learners.

With each passing year, many technological developments have improved the lives of students with special needs. It is well accepted that technological innovations and AI have played a significant role in interaction and communication. In terms of communication, the world has become flat, even as it supports the human aspect of interaction and communication difficulties. ASD is an impairment of basic social skills.

Though ISCs are still a work in progress, there have been various challenges urban design architects are recognising and addressing, which could result in inclusivity to all – both abled and disabled, both old and young – without discrimination. Use of e-readers, digital and multimedia technology enhances verbal and comprehension skills. There have also been wearable technology (WT), humanoids and various portable apps that have been used to enhance social skills though machine learning (ML) in individuals on the spectrum. Crowdsourcing and interactive maps have also been studied for the ease of movement within urban spaces.[2]

ASD is a condition where individuals have difficulties in social interaction and communication. These individuals show some repetitive or restricted behaviour and interests, lack language skills and have a limited range of resources that they can perform. People with ASD may display symptoms that hurt the person's ability to function properly in school, work and other areas of life. Estimations show that 0.625% of children are affected by ASD all over the world. However, with proper care and therapies, the functional aspects of children with autism can be improved (Elsabbagh, 2012).

The objective of this chapter is to identify the needs of the children with ASD from the parents' perspective. Essentially, each parent would wish well for their children and want to provide the best facilities for them. The quality function deployment (QFD) is used to understand the voice of the parents to understand what they seek from the future systems so that their children become a productive part of inclusive societies. This chapter has the following sections: Section 7.2 covers the literature review and highlights the literature gap. Section 7.3 discusses the mixed methods methodology. Section 7.4 provides qualitative insights from the parents' discussion across various age groups from different Indian cities. In Section 7.5, based on qualitative insights, the QFD method will first highlight what has to be concentrated on when developing the technology, thus enabling smart cities to create a system to ensure the quality of life for all.

7.2 LITERATURE REVIEW

In the extant literature, many researchers have considered the impact of technology in education. Moreover, many have considered the effect of this technology on the education and functioning of special needs learners, which further highlights research gaps and motivation. The literature highlights the current trends of technologies in smart cities and their challenges. In this section, we further discuss the benefits of AI in smart cities for people. Due to the restrictions on word limits, we have highlighted a few very recent relevant studies in this context. We initially, in general, share the development of smart cities and development and their mention in research.

One of the most recent studies by Radu (2020) explores the four disruptive technologies of Internet of Things (IoT), big data, AI and blockchain for the development of smart cities. The study illustrates their use for each specific smart city component: smart economy, smart mobility, smart governance, smart people, smart living and smart environment and further explores their opportunities and challenges. The chapter also highlights concerns such as loss of privacy, non-reversible negative environmental impact, the lack of attention on pertinent goals and problems related to integration into society. Nevertheless, disruptive technologies have been considered to be the key drivers when looking at smart cities. The success of these technologies necessitates public and private sector collaborations as well as government and citizen collaborations. Citizens need to be appraised about the benefits and risks of these technologies.

Cena et al. (2021) outline an app that was developed from the PIUMA (personalised interactive urban maps for autism), which aims to cater to individuals on the spectrum through personalisation and crowdsourcing. They recognise the limitations of the existing maps in urban spaces to serve the physically disabled population more than the cognitively impaired. This is an important project as this could really bring the smart cities project one step closer to being truly inclusive in nature.

Further, since the context of the paper concerns children with ASD, we highlight the details in the literature which highlight the relationship of the use of technology with ASD.

Porayska-Pomsta et al. (2018) have studied the efficacy of a learning environment with children diagnosed with ASD who engage in social interactions with an AI virtual agent and a human practitioner who acts in support of the exchanges. This chapter presents ECHOES, a design and evaluation technology that combines human and AI support for autistic children aged 4–14 years old, exploring their social interaction skills. ECHOES is a single-user technology–enhanced learning environment that utilises an artificially intelligent virtual character, called Andy, as a social partner for children with ASD and their typically developing peers, mainly to help them with their social communication skills. The analysis revealed a significant increase in the social responses made by ASD children to the human practitioners. The ASD children also exhibited increased interaction to both the virtual agent and the human practitioners throughout the sessions. However, this did not reach significance as there were significant individual differences within the ASD group. The study contributes significantly to the growing body of research in using AI technology design for autism intervention in real school contexts.

Further, in their study, Georgescu et al. (2019) adopted a proof-of-principle analysis of data from a social interaction study. This was done to train a classification algorithm on intrapersonal synchrony as an automatically and objectively measured phenotypic feature. The parameters considered were non-verbal motion energy values from 116 videos of social interactions. The classification approach has been applied to non-verbal behaviour objectively captured during naturalistic and complex interactions with an actual human interaction partner, assuring high external validity. The study concludes that an ML approach can capture diverse and complex behaviour in fundamental social interactions and will be essential in developing automatised and objective classification methods in ASD. Detailed literature by Song et al. (2019) covers one spanning from 2009 to 2019. The paper presents the studies which have considered AI technology about ASD. Some articles discussed how uses of AI technology for assessment in ASD screening and diagnosis give details of how AI has been able to enhance the diagnosis of ASD.

Xu and Geng (2019) proposed the theoretical and technical frameworks of people-centric service intelligence. This concept is very relevant to smart city research. In these frameworks, all needs are classified as six layers, namely, computational implementation, theory of behaviour, theory of attitude, theory of value, social theory and theory of self-actualisation according to Maslow's theory. The implementation techniques of smart cities are organised in levels of "digital city", "open city" and "intelligent city" consistent with the function divisions in computational implementation. The authors say that it will be helpful for both governments and industries to design people-centric service intelligence to improve quality of life, sustainability and for better development. The concept of people-centric service intelligence and its implementation routes could change the present circumstances of data intelligence to satisfy the human demands on urban functions. However, there are still some challenges in implementing people-centric service intelligence in the current stage. Multi- and cross-disciplines should be involved together to solve its challenges.

In another study, Al Banna et al. (2020) proposed an AI-based system that uses sensor data to monitor the patient's condition. The learning method can be adjusted through exciting games and tasks based on the person's facial expression. The proposed system consists of a smart wrist band (SWB), an interactive monitor and a camera device attached to the monitor for monitoring ASD patients. These devices will be connected to a mobile application to continuously monitor and keep them in a learning environment all day long without caregivers. The whole architecture will monitor the child's concentration, health condition and activity continuously. If it finds something unusual, it will send messages to the parents immediately. It will also generate a weekly report, including the child's daily activity, performance in games, mental state, activity log, progress in learning and health condition. The parents can easily find the growth of their child through it and send it to the doctor. This AI-based system has been of great help, especially during the COVID-19 pandemic, as ASD patients needed ongoing assistance, which was tough to get during the pandemic. The proposed automated AI-based control system will monitor the patients, support learning and therapy and help the parents adjust to the new situation.

A randomised controlled trial conducted in Holland Brookview Kids Rehabilitation Hospital in Toronto by Nguyen et al. (2021) indicated that an early detection of anxiety using a WT and being self-aware of these early signs could help individuals on the spectrum to implement calming strategies in stressful situations. The sample size for this study was of 28 individuals with ASD 8–18 years old, and the WT used was called Anxiety Meter. Over a span of three visits, the individuals selected were trained on how to use the Anxiety Meter as well as taught the diaphragmatic breathing technique to control anxiety. On their fourth visit, they were randomly assigned to either receive feedback with regard to their anxiety levels or not receive any feedback while completing anxiety-eliciting tasks. It was noted that real-time feedback did help most of the participants.

7.3 RESEARCH GAP AND MOTIVATION

Unfortunately, ISC surprisingly turns out to be under-investigated among academic and industrial communities. An ISC should follow a citizen-centric approach (Kamel Boulos et al., 2015) and put the welfare of citizens first (Joss et al., 2017; Maxwell, 2018) to cultivate social and ethical plurality, flexibility, creativity and open-mindedness in the ISC development process (Nam & Pardo, 2011a). In the present study, the Voice of Customers (VoCs) considered is the voice of the parents. Through the QFD approach we try to analyse the need for AI and technology intervention for the better management of children with autism.

From the extant literature, we can observe that the voice of parents is not considered in any of the papers, which highlights their requirements from the system for their children. Keeping in mind the need for such a study, owing to the customer-centric market, we develop this study to assess systems for ASD children using QFD. QFD helps determine the focus areas of improvements for a system based on the parents' opinions of ASD children.

7.4 METHODOLOGY

The methodology followed in this research is classified as a sequential exploratory mixed methods design. The research was conducted as per the two below-mentioned phases to quantify and generalise the qualitative data and generalise the qualitative data collected.

7.4.1 Phase 1: Qualitative Interviews with Parents of Children with Autism Spectrum Disorder

Extending the concept of quality improvement for developing the inclusive system of education and work for children with ASD, we deploy a two-step mixed methods approach in this research. In the first phase of this research, qualitative interviews are taken from ten parents to understand from the parents' perspective how they want the systems to develop to ensure an inclusive approach is designed for their children with ASD.

7.4.2 Phase 2: Qualitative Results Were Quantified using the Quality Function Deployment Approach (QFD)

We posted the detailed insights based on the parents' opinions in the interview; in the second phase, we were able to gather information in the voice of the parents for areas of need for development to ensure that the future overall systems are inclusive for their children. In this phase, we used QFD to quantify the qualitative results.

This structured approach gives increased importance to understanding customers' requirements and thus follows the customer empathy principle (Bouchereau & Rowlands, 2000). QFD is a methodology that helps to design a product or service based on the specific needs spelt out by the user of the product or service (Prasad and Chakraborty, 2013). The QFD methodology helps develop characteristics of a product or service that allow it to be considered high quality. QFD can also be very useful in determining how to deliver quality products and services in-line with customers' needs. In simple words, it works like a planning tool that begins with research to identify the VoC or, in other words, "what" the customers need (Hwarng and Teo, 2001).

QFD helps in identifying customer needs and their level of importance for the customer. It further helps transfer this understanding to decide design parameters for the product or service to achieve customer satisfaction (Eldin, 2002). One of the products of QFD is a "house of quality" (HOQ) matrix, which enables an easy-to-understand visual representation of "what customers want" versus "how it can be given to them" (Hauser and Clausing 1998).

Figure 7.1 gives in detail the overview of the methodology of research, including the steps of QFD.

Case analysis for application of QFD in decision-making:

Participants: The samples were the parents of ten children (two girls and eight boys) with autism residing in metro cities of India. This was a snowball sampling. Initial participants were known to the researchers, and the rest were the referrals given by the initial participants. The age group of children 12–19 years already diagnosed, parents developed acceptance, taking help of therapeutic interventions and going to regular schools. All of them belong to the nuclear family. All the parents are either homemakers or working in the private sector, and all are postgraduates belonging to the upper middle class. Participants were irrespective of the skill they possessed.

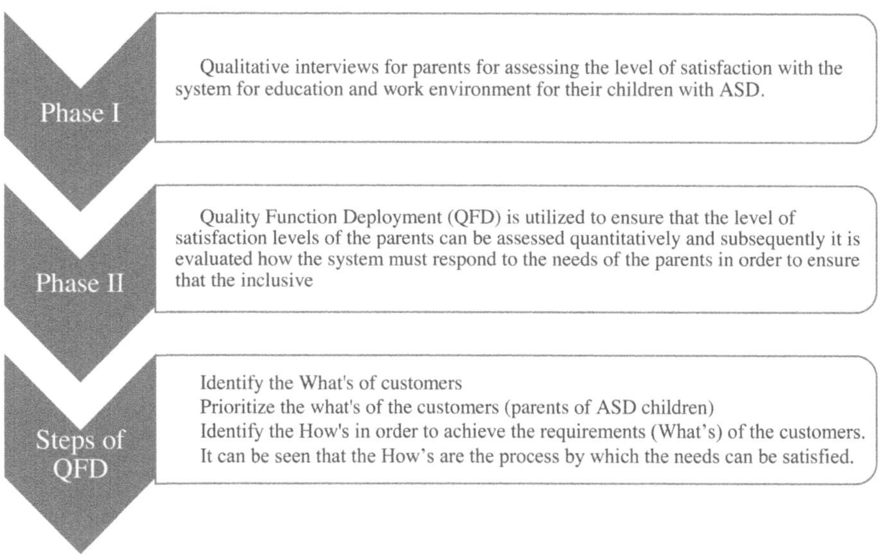

FIGURE 7.1 Methodology of research.

Procedure: Parents were interviewed through Google Meet for 45 minutes using a semi-structured questionnaire by a single researcher to control researcher bias. The needs of the parents were given to the participants to choose from, and they eventually ranked them according to their preferences. Few questions were open-ended, and the parents' responses were documented on an Excel sheet for further analysis.

Questionnaire: The research questionnaire was involved in exploring the technology intervention to satisfy the needs of the parents. As autism is a developmental disorder, our questionnaire focused on the difficulties faced and the needs of the children considering their educational, home and other environmental set-ups. This led us to divide the questionnaire into two parts: what the needs of the parents are and how the needs are satisfied by using AI and other technology interventions. The specific questions in each section were suggested based on our clinical and therapeutic experience working with children with autism.

Data analysis: All verbatim observation transcripts and interviews were scanned for "what are the most important needs of their child" and "what features of a technology-based smart city will support the child's needs". All data gathered were analysed for concepts, context and process using the constant comparative method (Corbin & Strauss, 2007). Each need was coded and continually compared with other instances in the data for similarities and differences. By continually comparing specific references in the data, investigators refine identified concepts, explain their properties and explore their relationship to one another. Once themes emerged, we quantified the data using the house of quality matrix and the QFD approach. Findings were integrated to form tentative explanations of the needs of the parents who have children with autism that require technological intervention to be included as part of a smart city.

7.5 RESULT AND DISCUSSION

Common themes emerged in the data across all participating parents who have children with ASD. Participants represented the same range of age and grade levels; the contexts in which they perceived needs are somewhat similar. Figure 7.2 highlights the parents' needs that they have emphasised that they need from an ISC for their children with ASD, and this highlights characteristics they want as features of the ISC.

Once the parents highlight the need, these must be converted into technical characteristics of the system. These technical characteristics, terms as the "how's", will essentially help achieve the requirements of the parents. The technical characteristics were discussed with the parents, subject matter experts and highlights from the extant literature. Figure 7.3 highlights the technical aspects.

The House of Quality matrix is given in Table 7.1. In this table, the strength of the relationship between the technical characteristics (how's) is given with regard to each What. The relationship of each technical characteristic on the what's was assessed based on the 1, 3, 9 scale as suggested by Raharjo et al. (2008). The importance

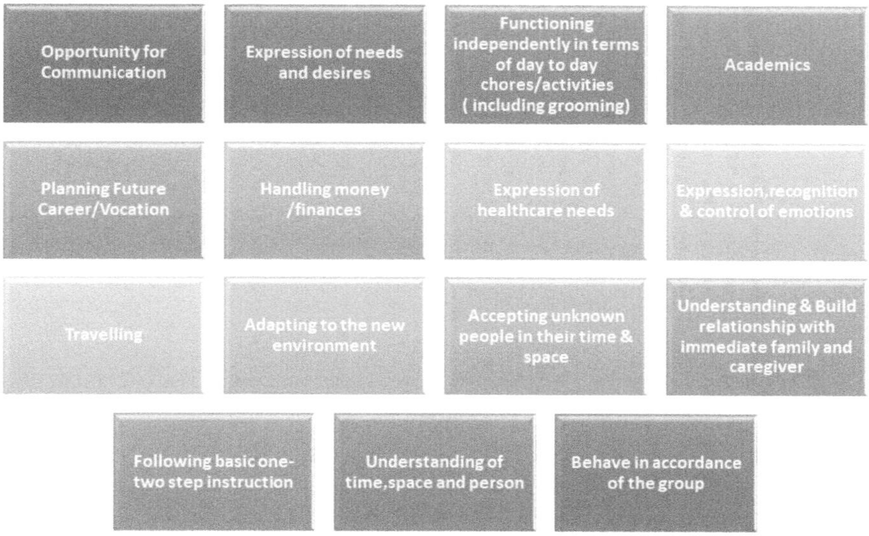

FIGURE 7.2 Needs highlighted by parents for ISC.

ranking of the what's was also calculated based on responses from the parents in the interviews. The combined weighted average value for each technical characteristic based on what's and the ranking of the technical characteristics (how's) gives us the final importance level of characteristics. Based on the importance of the characteristics, it provides the detail of what should be the focus of educational institutions and industries in the development of ISC to cater to the needs of the children with ASD as desired by their parents.

FIGURE 7.3 Response of parents for technical characteristics (how's).

TABLE 7.1
House of Quality Matrix

	Weight	Classrooms with Audio-Visual Support	Availability of Equipment to Satisfy the Kinesthetic Needs of the Child	Auditory Aid to Adjust the Sound According to Level of Tolerance	Availability of Devices with Accessibility Tools for Reading and Writing	Technology-Enabled Non-judgmental Communication (Robots, Bots, Alexa, Siri)	Lessons with Augmented Reality/3D Printing	Applications for Reminders/ Schedules	Opportunity for Talent Management through Technology-Enabled Apps	Smart Cards for Travelling	Google Maps for Directions	Technological Intervention for Awareness Building for Others	Instructions in Public Places to be Suitable for the Child (Repeated at a Proper Pace with Visual Clues)	Simulations for Group Behaviour (Representing different situations/ people)
Opportunity for communication	5	8	3	8	1	8	1	1	3	3	0	3	9	3
Expression of needs and desires	4	3	1	3	3	8	1	3	3	1	0	1	9	3
Functioning independently in terms of dag-to-dag chores/activities (including grooming)	4	0	9	8	9	8	1	8	3	9	9	1	9	3
Academics	3	8	3	3	9	3	8	3	3	0	0	0	3	3
Planning future Career/Vocation	4	3	1	1	3	1	8	3	3	0	0	1	3	3
Handling money/ finances	3	0	3	3	9	1	3	3	8	1	1	0	3	3
Expression of healthcare needs	4	0	0	8	9	3	0	8	3	0	0	0	0	0
Expression, recognition and control of emotions	2	3	3	3	3	8	0	0	8	0	0	3	3	9

(Continued)

TABLE 7.1 *(Continued)*
House of Quality Matrix

	Weight	Classrooms with Audio-Visual Support	Availability of Equipment to Satisfy the Kinesthetic Needs of the Child	Auditory Aid to Adjust the Sound According to Level of Tolerance	Availability of Devices with Accessibility Tools for Reading and Writing	Technology-Enabled Non-judgmental Communication (Robots, Bots, Alexa, Siri)	Lessons with Augmented Reality/3D Printing	Applications for Reminders/ Schedules	Opportunity for Talent Management through Technology-Enabled Apps	Smart Cards for Travelling	Google Maps for Directions	Technological Intervention for Awareness Building for Others	Instructions in Public Places to be Suitable for the Child (Repeated at a Proper Pace with Visual Clues)	Simulations for Group Behaviour (Representing different situations/ people)
Travelling	3	0	3	3	3	8	1	3	1	9	8	1	9	3
Adapting to the new environment	3	0	1	1	0	8	0	3	8	0	1	3	1	9
Accepting unknown people in their time and space	2	0	0	0	0	8	0	1	3	0	0	8	3	9
Understanding and building relationship with immediate family and caregiver	2	0	1	3	1	3	0	3	3	0	0	3	0	9
Following basic one-two step instruction	1	3	0	3	3	3	1	3	3	0	0	1	3	3
Understanding of time, space and person	1	1	3	3	3	3	1	3	3	0	0	1	9	9
Behave in accordance with the group	5	1	1	3	1	8	3	0	8	0	0	9	1	9
Weighted average			105	196	183	282			210	85			208	216

TABLE 7.2
Top Five How's

Technology-enabled non-judgmental communication (Robots, Bots, Alexa, Siri)	292
Simulations for group behaviour (representing different situations/people)	216
Opportunities for talent management through technology-enabled apps	210
Instructions in public places to be suitable for the child (repeated at a proper pace with visual clues)	206
Auditory aid to adjust the sound according to the level of tolerance	196

Table 7.2 highlights the top 5 how's, which should be the focus area to develop the ISC for children with ASD.

7.6 CONCLUSION

From the study, it can be concluded that there is prominent need to have a 'critical focus' by parents. The top five need via QFD are Technology-Enabled Communication, need of Simulation during group interactions, Talent Management through digital apps, suitable Instructions given to child (along with visual clues) at public places and Auditory or Reading or Writing Aid to adjust the level of the sound. It can be observed that autistic individuals are a primary significant concern in smart city concepts too. Smart cities should have all the facilities for autistic people to live everyday life without the assistance of others and enjoy all the modern amenities of the newest technology, which will eventually make their lives easier. Designing technologies and devices that have AI can lead towards the goal of a smart city. The ML- and AI-enabled devices can help them learn by evaluating their condition and keeping them within a controlled environment without the need for external support. The limitation of the study is that it has only considered the perspective of parents in the development of the structure of smart cities for ASD. It is essential to also take into consideration the opinion of the other stakeholders as well. The limitations of this study would lead to the future scope of the study.

NOTES

1. https://www.cmhc-schl.gc.ca/en/nhs/nhs-project-profiles/2020-nhs-projects/housing-through-an-autism-lens
2. https://www.nlc.org/resource/smart-city-development/

REFERENCES

Albino, V., Berardi, U., & Dangelico, R. M. (2015). Smart cities: Definitions, dimensions, performance, and initiatives. Journal of Urban Technology, 22(1), 3–21. https://doi.org/10.1080/10630732.2014.942092.

AlAwadhi, S., & Scholl, H. J. (2013). Aspirations and realizations: The Smart City of Seattle. 2013 46th Hawaii International Conference on System Sciences (pp. 1695–1703). IEEE. https://doi.org/10.1109/HICSS.2013.102.

Al Banna, M. H., Ghosh, T., Taher, K., Kaiser, M. S., & Mahmud, M. (2020). A monitoring system for patients of autism spectrum disorder using artificial intelligence. In Mahmud, M., Vassanelli, S., Kaiser, M.S., Zhong, N. (eds.), Brain informatics. Vol. 12241. Springer, Cham. https://doi.org/10.1007/978-3-030-59277-6_23

Bouchereau, V., & Rowlands, H. (2000). Methods and techniques to help quality function deployment (QFD). Benchmarking: An International Journal. https://ieeexplore.ieee.org/document/6480046

Cena, F., Rapp, A., Mattutino, C., Mauro, N., Ardissono, L., Cuccurullo, S., ... & Tirassa, M. (2021). A personalised interactive mobile app for people with autism spectrum disorder. In Ardito, C., et al. (eds.), Human-computer interaction. INTERACT 2021. Vol. 12936. Springer, Cham. https://doi.org/10.1007/978-3-030-85607-6_28

Clark, M., & Adams, D. (2020, August 13). Listening to parents to understand their priorities for autism research. PLoS One. Retrieved from https://journals.plos.org/plosone/article?id=10.1371%2Fjournal.pone.0237376

Corbin, J., & Strauss, A. (2007). Basics of qualitative research (3rd ed.). Thousand Oaks, CA: Sage.

de Oliveira Neto, J. o. S. (2018). Inclusive smart cities: theory and tools to improve the experience of people with disabilities in urban spaces. Université Paris-Saclay.

De Santis, R., Fasano, A., Mignolli, N., & Villa, A. (2014, March 1). Smart city: fact and fiction. Retrieved from http://mpra.ub.uni-muenchen.de/54536/1/MPRA_paper_54536.pdf

Eldin, N. (2002). A promising planning tool: quality function deployment. Cost Engineering, 44(3), 28.

Elsabbagh, M., et al. (2012). Global prevalence of autism and other pervasive developmental disorders. Autism Research, 5(3), 160–179.

Fernandez-Anez, V. (2016). Stakeholders approach smart cities: A survey on smart city definitions. Lecture Notes in Computer Science, 9704, 157–167. https://doi.org/10.1007/978-3-319-39595-1_16

Fernández-Güell, J.-M., Collado-Lara, M., Guzmán-Araña, S., & Fernández-Añez, V. (2016). Incorporating a systemic and foresight approach into smart city initiatives: The case of Spanish cities. Journal of Urban Technology, 1–25. https://doi.org/10.1080/10630732.2016.1164441

Georgescu, A. L., Koehler, J. C., Weiske, J., Vogeley, K., Koutsouleris, N., & Falter-Wagner, C. (2019). Machine learning to study social interaction difficulties in ASD. Frontiers in Robotics and AI, 6, 132. doi: 10.3389/frobt.2019.00132 https://www.frontiersin.org/articles/10.3389/frobt.2019.00132/full

Hauser, J. R., & Clausing, D., 1998. The house of quality. Harvard Business Review 76 (3), 63–73.

Hwarng, H. B., & Teo, C. (2001). Translating customers' voices into operations requirements – A QFD application in higher education. International Journal of Quality & Reliability Management. https://doi.org/10.1108/02656710110379075

Joss, S., Cook, M., & Dayot, Y. (2017). Smart cities: Towards a new citizenship regime? A discourse analysis of the British smart city standard. Journal of Urban Technology, 24(4), 29–49.

Kamel Boulos, M. N., Tsouros, A. D., & Holopainen, A. (2015). 'Social, innovative and smart cities are happy and resilient': insights from the WHO EURO 2014 International Healthy Cities Conference. International Journal of Health Geographics, 14(1), 3. https://doi.org/10.1186/1476-072x-14-3

Maxwell, L. (2018). How to ensure that your smart city strategy is inclusive smart city Strategy: Facilitating successful urban transformation.

Mattoni, B., Gugliermetti, F., & Bisegna, F. (2015). A multilevel method to assess and design the renovation and integration of smart cities. Sustainable Cities and Society, 15, 105–119.

Meijer, A., & Bolívar, M. P. R. (2016). Governing the smart city: a review of the literature on smart urban governance. International Review of Administrative Sciences, 82(2), 392–408.

Nam, T., & Pardo, T. A. (2011a). Conceptualizing smart city with dimensions of technology, people, and institutions. 282. New York, NY: ACM Press. https://doi.org/10.1145/2037556.2037602

Nam, T., & Pardo, T. A. (2011b). Smart city as urban innovation: Focusing on management, policy, and context. Retrieved February 10, 2015, from http://www.ctg.albany.edu/publications/journals/icegov_2011_smartcity

Nguyen, J., Cardy, R. E., Anagnostou, E. et al. (2021). Examining the effect of a wearable, anxiety detection technology on improving the awareness of anxiety signs in autism spectrum disorder: a pilot randomized controlled trial. Molecular Autism 12, 72. https://doi.org/10.1186/s13229-021-00477-z

Porayska-Pomsta, K., Alcorn, A. M., Avramides, K., Beale, S., Bernardini, S., Foster, M. E., Frauenberger, C., Good, J., Guldberg, K., Keay-Bright, W. (2018). Blending human and artificial intelligence to support autistic children's social communication skills. ACM Transactions on Computer-Human Interaction, 25, 1–35.

Prasad, K., & Chakraborty, S. (2013). A quality function deployment-based model for materials selection. Materials & Design, 49, 525–535.

Radu, L.-D. (2020). Disruptive technologies in smart cities: A survey on current trends and challenges. Smart Cities, 3(3), 1022–1038. doi:10.3390/smartcities3030051

Raharjo, H., Brombacher, A. C., & Xie, M. (2008). Dealing with subjectivity in early product design phase: A systematic approach to exploit quality function deployment potentials. Computers & Industrial Engineering, 55(1), 253–278.

Russell, S.J., Norvig, P. (2003). Artificial intelligence: A modern approach (2nd ed.). Pearson Education, Inc., Upper Saddle River, New Jersey.

Song, D. Y., Kim, S. Y., Bong, G., Kim, J. M., & Yoo, H. J. (2019). The use of artificial intelligence in screening and diagnosis of autism spectrum disorder: A literature review. Soa–ch'ongsonyon chongsin uihak = Journal of Child & Adolescent Psychiatry, 30(4), 145–152. https://doi.org/10.5765/jkacap.190027

Xu, H., & Geng, X. (2019). People-centric service intelligence for smart cities. Smart Cities, 2(2), 135–152. doi:10.3390/smartcities2020010

8 Innovation Management in the Context of Smart Cities' Digital Transformation

Pankaj Tiwari[1], *Pooja Ahuja*[1], *Gobinath Ravindran*[2], *and Seyed Jalaleddin Mousavirad*[3]
[1]SVN University, Sagar, India
[2]SR University, Hyderabad, India
[3]University of Beira Interior, Covilhã, Portugal

8.1 INTRODUCTION

Currently, around six billion people are residing in cities, which means a global trend of digital-based urbanization is occurring. An increasing number of metropolises are developing their smart city development efforts. This movement has led to global investments and policy innovations for technological implementations and data consumption to address social issues and urban growth (Anthony, 2018). Respectively, a smart city is a city that is able to generate, collect and process data to facilitate intelligent predictive and decisions analysis for better urban planning and development (Tomičić et al., 2019). According to the United Nations (UN) definition, a smart city is an innovative city that deploys Information and Communication Technology (ICT) and other mediums to enhance quality of life, efficacy of urban services and operation and competitiveness, while ensuring that it addresses the needs of present and future generations in regard to social, economic and environmental aspects. Smart city development is based on digital innovations provided by enterprises which provide smart services to improve citizens' quality of life (Anthony, 2018).

Smart cities include the creative utilization of digitalization which contains different types of ICT that are mixed to offer keen types of assistance (Goerzig & Bauernhansl, 2018). A brilliant city is a social design that unites business, society and innovation. For a city to be brilliant, the mechanical as well as the natural, social and human perspectives ought to be thought of (Mendhurwar & Mishra, 2019). Brilliant urban communities are upheld by digitalization and mechanical advancements that achieve future ecological, social and monetary advantages (Bokolo & Petersen, 2019). These carefully empowered urban areas are upheld by ICT, alluded to as computerized advances, which logically guarantee colossal possibilities for development. By and by, advanced change is perhaps the most recent improvement driven by mechanical advancement, which disturbs existing plans of action in various areas (Gampfer, 2018). Computerized

DOI:10.1201/9781003225317-8

change includes applying advances to a few areas of an endeavour, which lead to significant changes in undertakings and the manner in which esteem is created for partners. Computerized change intends to re-adjust cycles, innovation and plans of action to make an incentive for clients just as undertakings (Vobugari, 2017). Computerized change can achieve expanded efficiency and income and lower expenses, further developing customer administration (Aliee et al., 2019). Computerized change involves the association of entertainers over the worth chain and the arrangement of frameworks for social occasion, trade, handling and investigation of city information to help direction (Petrova-Antonova & Ilieva, 2018). Computerized change includes the use of advances to further develop city execution and scale tasks, administration and authoritative constructions (Goerzig & Bauernhansl, 2018).

As of now, urban areas are confronted with expanded strain to ceaselessly create a coordinated and quick evolving world. To effectively adjust in such conditions, urban areas are relied upon to conform to evolving conditions (Babar & Yu, 2015). In this manner, computerized change empowers individuals and independent gadgets to collaborate past their own setting using Information Technology (IT), working with huge information, distributed computing and portable and social advances (Gampfer, 2018; Such a change is a significant shift from the past business as usual and results in conceivably problematic metropolitan wide change, empowering districts to move from traditional activity to advanced methodology (Babar & Yu, 2015). Changing urban communities is a reaction to computerized change, which requires an organized methodology, especially in urban communities, that involves various substances with various mechanical and social components, all of which can administer the achievement or disappointment of advanced change (Babar & Yu, 2015).

8.2 SMART CITIES: THE TECHNOLOGICAL REVOLUTION REACHES THE CITIES

Feasible urban areas, which are associated with improved innovation, are not just a relentless worldwide peculiarity, but they are the main answer for containing and diminishing the disturbing ecological and financial repercussions that urbanization will have for our planet.

Consistently, more than 180,000 individuals move to a city to live. The Organization for Economic Cooperation and Development (OECD) figures that by 2050 the total world population will arrive at nine billion, of whom 70% will live in metropolitan communities. With cities previously burning through more than 75% of the world's energy creation and producing 80% of ozone depleting substance (GHG) emanations, numerous urban communities have chosen to deliberately reuse and carefully make changes because of noteworthy worldwide difficulties: population density increase, contamination, shortage of assets, water assets and energy effectiveness.

What's more, they have done this by becoming savvy urban areas or urban communities 4.0: they depend on ICTs and enormous information to adequately and reasonably oversee everything from transportation to the use of energy or water assets, public spaces and correspondence with their occupants. The objective: to lessen energy consumption, diminish CO_2 emanations and increase the prosperity of their occupants.

8.3 INCREASING DEMAND FOR SMART CITIES

The greater part the total population as of now consumes metropolitan spaces. Evaluations figure that the number will arrive at 66% by 2050. This advancement will occur eventually because of the numerous chances given to individuals that are granted a chance to plan their own lives in urban areas. Rising urbanization likewise implies more prominent difficulties: as urban areas develop individuals' necessities, requirements should be met in a manner that also back off the climate.

Environmental change is quite possibly the most major problem we are presently confronted with. CO_2 discharges should be decreased in a very short time, so measures should be taken to avoid facing a dangerous atmospheric deviation, floods and tremendous hot waves. Urban communities are responsible for roughly three-fourths of ozone-harming substances around the world. Instead of being significant polluters, they should be strictly called upon to provide tough measures.

Overall systems' administration of workforces, foundations and data has repercussions on urban communities as well. Financial and social constructions are changing, and metropolitan governmental issues need to adjust their systems to these new conditions. This implies situating urban communities universally among collaboration and rivalry. The actions taken should not fulfil the sole need of appearances whatever should zero be on inner social, financial, spatial and underlying perspectives also.

Savvy economies effectively support schooling, capability, research, advancement, usefulness and adaptability. Nonstop information procurement and move, just as nearby and worldwide organizations, are the primary elements for inventive yield. Ventures offering IT, natural and energy benefits specifically are viewed as the main thrust for brilliant economies.

Lessening energy, natural substance utilization and forward-looking assets are among a city's main pressing issues. Savvy supply and removal frameworks are similarly pretty much as significant as cycle-driven changes, innovative turns of events and organizations for energy, versatility, foundation and structures. Brilliant frameworks are a stage towards savvy energy consumption: astute organizations and checking frameworks are placed accountable for energy age, stockpiling and use. Shrewd meters are introduced to make real energy consumption in a more straightforward way.

8.3.1 The Smart City Management Model

The SCMM tends to the primary components of any smart city project:

- The administrations' regions, in the focal point of the chart, sum up the sort of drives embraced in a smart city project.
- The change cycle process, portrayed as a circle, shows the essential advances expected to succeed when fabricating a smart city programme: start change, decide the area, foster ideas and synchronize accomplices, actuate assets, acknowledge ventures and activity and instrumentation.
- The crucial components, around the edge, incorporate the two past focuses: information administration, accomplice and city support, financing, innovation, plans of action, correspondence.

- The SCMM is a mind-boggling system (smart cities are complicated, for sure). In any case, it merits assessing the essential components as they are the foundation of each effective venture in the field.

8.3.2 There Are Two Concepts that Should Be Employed with Respect to New Technologies in Smart Cities

1. **Absorptive capacity**
2. **Open Innovation**

8.3.2.1 Absorptive Capacity

Absorptive capacity is a term first used in business administration to refer to 'a firm's ability to recognize the value of new information, assimilate it and apply it to commercial ends'. In the context of smart cities and innovation, it means more or less the same: developing, anticipating and maintaining the absorption capacity are critical for the survival of the model.

The value of each innovation depends on our capacity to absorb and take the advantage of it. New technologies alone, without any social impact, don't actually have a value.

8.3.2.2 Open Innovation

Open innovation is something very tightly connected to absorption capacity. It means sourcing external knowledge for one's innovation management strategies. In a smart city context, we need open data for open innovation. In an ideal scenario, the institutions acknowledge crowdsourcing as a tool to improve the relationship to their citizens by integrating them into political decision-making.

8.3.3 Data Governance

A smart city is about data, a great deal of it. The computerized shadow of a city will deal with an enormous measure of data consistently. Additionally, it has likewise to be accessible for accomplices, both public establishments and people, who will work with that crude information the administrations require for the various regions including, subsequently, the significance of information administration.

Legitimate information administration ought to manage how information is gathered, dealt with and shared. It incorporates information protection in execution, a basic issue that shouldn't be excused, just as information unwavering quality. This point is vital as smart cities are digitizing information from fundamental administrations, like energy, public transportation, water supply or broadcast communications. Urban communities can't bear the cost of helpless information for execution with these administrations!

8.4 CHARACTERISTICS OF DIGITAL TRANSFORMATION

The issue of digitization is regularly connected with improving business measures. In this specific situation, computerization and normalization are key segments in the chase for further developed proficiency in some random business. Basically, it focuses on the subject of executing less cost-productive cycles and techniques. By taking these actions, organizations endeavour towards advancing their market position, whereas firms coming up short on the abilities or inspiration to develop their methods of working together will in general end up at the rear of the pack. Indeed, digitization endeavours are frequently seen as awkward and require huge assets; however, the prizes of better client-driven activities and simpler adaptability ought to be considered. One of the apparent benefits of digitization is admittance to business and market information, which can be handled and used for further developing vital dynamics. In the light of a VUCA business climate, complex information, the executives framework, can turn into a significant factor for future achievement. Thus, firms will in general be zero in all-encompassing programming choices as a method for supporting their digitization endeavours. Considering all, firms have been faced with those difficulties for more than 20 years. Shockingly however, just a predetermined number of organizations have been prepared to do effectively executing normalized and mechanized cycles. As indicated by research directed by MIT, approximately one of every four set-up firms has completely digitized their inward methodology and cycles. This is especially disturbing given that digitization endeavours are viewed as an initial move towards accomplishing the benefits of transforming computerized change into new business openings (Markovitch & Willmott, 2014; Nandram & Bindlish, 2017; Ross, 2017).

Against the foundation of advanced change, organizations face significant difficulties with innovative improvement, for example, distributed computing, web of things and computerized reasoning resetting the guidelines of the game in business. Given a large number of points descending upon organizations, they experience issues throughout in going through this intricacy. Moreover, changing methods of working together under these conditions can be attempted for singular market players who have been choosing a lease looking for modus for an all-inclusive timeframe. Computerized change flips around business sectors and their players with the principles of market interest can be reclassified. As an outcome, the force elements among firms and clients is moved, leaving organizations with open issues in the field of making new incentives for their customers, drawing in computerized sagacious ability, assembling and keeping a culture of sharing and minding among others. Compactly, advanced change come a long way from past, building other application. It comes down to carrying out another method of working together (Schallmo et al., 2016; Ross, 2017; Newman, 2018; Richards, 2018).

To work with the advanced change venture, Futurum Research (2018) distributed a yearly computerized change record to lay out the most fundamental effect factors for inside driving turns of events and changes to accept the unfolding advanced time. In this specific circumstance, Figure 8.1 underlines the significance of innovative organizations; in blend with a steady organization culture, including the board also to help gather gifted workers. This basically sums up the foundation of computerized astute organization accepting change and development. The following figure (Figure 8.2) conveys about the digital transformation and the internal drivers that persuade it.

8.4.1 Digital Transformation and Digital Platforms

Digital transformation of social and financial collaborations is typified in the multi-sided advanced stage, which is a more significant class for understanding computerized change than the Internet, programming and blockchain (Vial, 2019). In view of

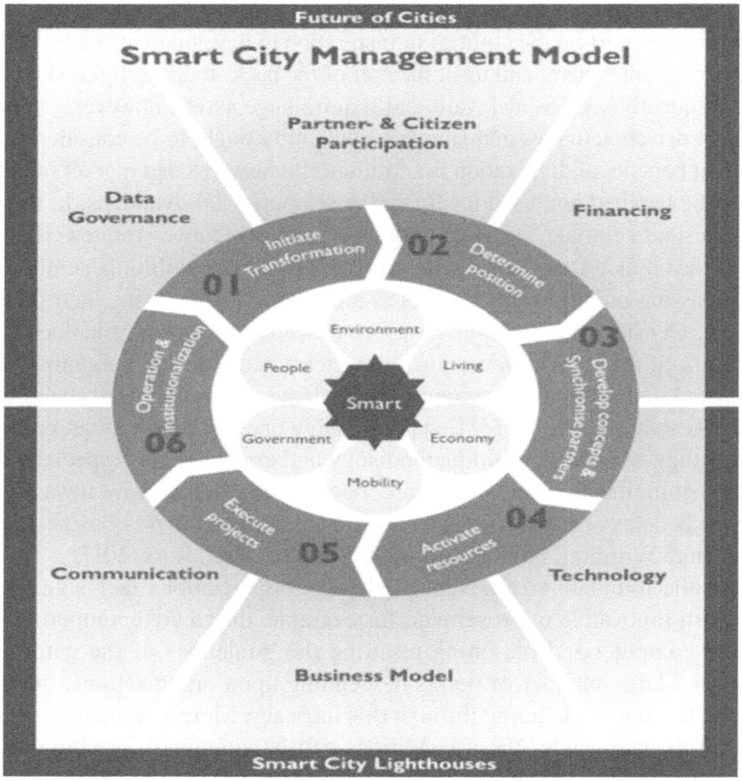

FIGURE 8.1 On the off chance that smart cities are so intricate, an inquiry may emerge in the brain of metropolitan designers: Is there an aide for that? Indeed, there is: the smart city management model (or SCMM). It depends on research led at the University of St Gallen by the creators. The SCMM is shown underneath.

Hänninen et al. (2017), one might say that the stage economy is going through a computerized change of significant worth creation (Hänninen et al., 2017). Investigations of advanced business change transcendently centre around money, showcasing and development of the board (Hausberg et al., 2019), whereas investigations of the development of business and stage environments centre around the collaboration of related entertainers in these biological systems (Hermes et al., 2020) or on portraying the historical backdrop of individual stages as far as business process rationale rather than the rationale of computerized change (Skog et al., 2018). Hanafizadeh et al. (2020) recognized four hypothetical structures in the data frameworks field – innovation dissemination, computerized partition, e-consideration, innovation reception and use, and participative plan – and exhibit their mistakes for the policymaking and administration of advanced stages. As opposed to past advances, including data, computerized stages are multisided and moulded by communications, which have a layered design. Be that as it may, Hanafizadeh et al., in their review for the improvement of computerized stage hypothesis, didn't focus on the effect of digitization collaborations on stage highlights.

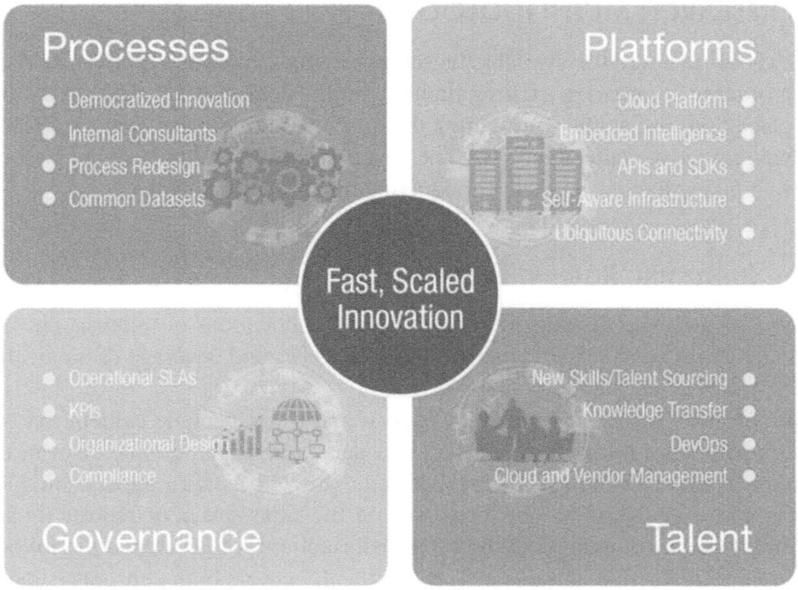

FIGURE 8.2 Internal drivers of digital transformation.

Hence, there is an absence of exploration in the logical writing looking at the entrapment between the advancement of computerized stages and the cycles of digitization/digitalization/computerized change. Simultaneously, it is pivotal for understanding the particularity of advanced change to recognize digitization processes dependent on I3.0 and I4.0 innovations, individually.

8.4.2 Data and Innovation

The common feature of many disruptive technologies – and industries – is big data. Abundant personal and professional data can be harvested from 'always on' mobile devices and used to drive anything from real-time service adaptations to high-level decision-making and new product design.

As per the reports, by 2022, 53% of IT projects will create new digital services and revenue streams that monetize data. In the customer experience realm, transformative projects will include analytics-based customer segmentation, predictive marketing, personalization and enhancement of touch points to align with customer journeys. On the operational side, there will be a greater focus on the workforce with data-driven decision-making and, ultimately, data-informed cognitive analytics.

Finally, although only 40% of consumers and retail brand executives say they are using data and analytics to strategic ends today, according to KPMG's 2019 Global Consumer Executive Top of Mind Survey, the number is expected to be increased to 63% over the years. Moreover, predictive analytics are projected to increase it from 26% to 53% in the same time period.

8.5 RESEARCH METHODOLOGY

This chapter reviewed the available literature on smart cities and digital transformation in order to summarize existing findings in this field of study. Special focus has been placed on case studies, presenting the initiatives in practice, applying quantitative and qualitative content analysis. The next two subsections present the methodology of choosing the literature and conducting the literature analysis.

8.5.1 Identifying Relevant Literature

Literature search was conducted in order to find publications of interest. Three scientific databases were selected due to their relatively good coverage of publications: IEEE, Web of Science and Scopus.

The selection was done on all 134 papers, and it has become evident that some of the papers were found in more than one database, so the double or even triple appearances had to be checked and resolved. This left us with papers suitable for our qualitative analysis. As our focus was on the experiences or reports on smart citiy initiatives in the context of digital transformation, we developed the following criteria for the selection of papers: (1) it specifically covers both concepts (smart cities, digital transformation) and (2) it reports on a case study, initiative or a project.

8.5.2 Review Analysis

In order to achieve the aim of the article, all relevant data (e.g. authors, title, publication source, year of publication, scope with subsections, services with examples, technology/technologies implemented) from the selected papers have been extracted.

Next, a framework for the study has been developed, as described further in the text. Our first findings show that there is a lack of research on the topic, although it is a growing and developing area. In order to explore the reported experiences or reports on smart city initiatives in the context of digital transformation, we generated a conceptual framework based on the observation of the selected 42 papers. The proposed conceptual framework suggests three dimensions:

1. **Scope of the smart city case study** – it describes strategic alignment of transformation projects and initiatives to specific goals, aims and purpose. The scope is indicative to recognizing beneficiaries in case it is not explicitly stated who the targeted audience is.
2. **Services that were planned or delivered in the case study** – this concept is related to specific output, which is planned for an initiative or a project. This dimension is also indicative to technologies implemented by the initiative.
3. **Technology that was planned or used to generate service** – with a focus on technologies related to Industry 4.0.

Within these dimensions, there are various sub-concepts. The initial list of sub-concepts was formed by gathering scope, service and technology concepts from various sources, whereby the most resembling research showed that these categories were described as aims/goals or themes (Ismagilova et al., 2019). However, for the purpose of our analysis, the three main concepts are selected scope, service and technology dimensions, and the sub-concepts have been adapted to fit the selected case studies' contents. In addition, sub-concepts have been coded by the authors for the purpose of the analysis of scope, service and technology.

8.6 TECHNOLOGY-DRIVEN INNOVATIONS IN SMART CITIES

Technology plays a vital role in smart cities, and innovative technological strategies efficiently help cities become smart. Such innovations often cannot be restricted at a process level, and a change in the overall model of operations is often required. Sometimes innovative solutions reduce time and effort by government, and the plan is executed safely.

Two technology-driven solutions that can help governments implement smart cities are discussed in this chapter: (1) 'Leveraging Smart Supply Chain and Information System Agility for Supply Chain Flexibility' by Gupta et al. (2019). They employed organization information processing theory to achieve greater flexibility in the supply chain between customer and supplier. The data for this study have been collected through a survey on a Likert scale from 150 respondents working in South Africa. The authors used structured equation modelling for data analysis of the survey data. The results suggested that, in comparison to the customer, suppliers have more effect on the relationship between information system agility and supply chain flexibility. A supply chain with information processing capabilities may hamper supply chain flexibility due to the delay caused by the centralized information processing system.

(2) The study by proposed a model based on individual as well as organizational technology adoption in the article '"Farmer Cooperatives" Intention to Adopt Agricultural Information Technology—Mediating Effects of Attitudes'. Two widely used frameworks were integrated, which are the technology acceptance model (TAM) and the technology-organization-environment (TOE) framework. The 288 respondents' responses on the data through a questionnaire floated in and around China, and Confirmatory Factor Analysis (CFA) was analysed to check the validity and reliability of the proposed model. It was concluded that environmental factors play a more important role for the adoption of agricultural ICTs in comparison to the technological and cooperative focused factors.

8.6.1 Datafication

The data-driven city executes datafication for upgrading and streamlining its tasks, capacities, administrations, procedures and arrangements. Comprehensively, datafication alludes to the aggregate instruments, processes, strategies, procedures and advancements used to change a city to an information-driven venture. The increase of datafication is shown in the extreme development in the volume, reach, assortment and granularity of the information produced about metropolitan conditions and residents (Bibri & Krogstie, 2021; Crawford & Schultz 2014; Kitchin 2014, 2016;), with the mean to evaluate the various parts of urbanity in the advanced city.

Urban communities today are reliant upon their information to work appropriately—and even to work at all with respect to numerous areas of metropolitan life. A city that executes datafication is supposed to be datafied. To datafy a city is to place it in a measured configuration with the goal that it tends to be organized, saddled and dissected. Urban areas are presently taking any conceivable quantifiable measurement and extracting esteem from it by upgrading independent directions relating to a wide assortment of pragmatic uses comparable to numerous metropolitan frameworks and spaces. In cutting edge information situated in a metropolitan scene, a city's presentation is estimated, evaluated and improved depending on the capacity of having command over the capacity, execution, handling and examination of the metropolitan information, just as on the information removed from these information as applied metropolitan knowledge. Handling the difficulties of supportability and relieving the adverse consequences of urbanization are among the vital worries of the datafication of the cutting-edge city (Bibri & Krogstie, 2021).

8.6.2 Data Analytics Is Playing a Vital Role in Urban Planning or the Development of Smart Cities

Developing public consciousness of ecological supportability, fossil fuel by-products and an assortment of different issues is making legislatures perceive the need of making individuals focused urban areas, that is, urban communities planned around residents rather than vehicles or organizations. These 'brilliant urban areas' are considered to address a future pattern in metropolitan arranging and are fundamental for handling metropolitan issues. Information examination is a major piece of this new plan pattern. Information examination assumes a basic part in helping urban areas become carbon-impartial, working on metropolitan transportation and dealing with their framework in a protected, long-term and practical way. Information investigation is a pivotal part in keen city arranging.

The rising number of Internet of Things (IoT) gadgets circulated all through the metropolitan climate can possibly modify our view of how urban communities work. Huge information investigation, artificial intelligence (AI), computerized reasoning and information perception are largely accelerating, giving city heads and pioneers devices to assist in detecting these information streams. All the while, urban areas are assembling new information reconciliation and sharing stages, which are separating ordinary functional storehouses and permitting all partners and planned arrangement providers access.

Subsequently, large information's benefits are a basic part of many shrewd city drives. Prescient investigation of traffic and transportation designs, for instance, can lessen clogging and work on the productivity of public transportation administrations, city assets for public security, social consideration, and other key administrations can be designated all the more viably utilizing around date examination, energy proficiency projects can be aimed at the most weak families and structures, reasonable for retrofit projects and open information stages can build resident commitment and energize novel thoughts.

All components of public help and city in execution can profit from information examination. Coming up next are a few models. The progress to identify carbon urban areas: advanced information investigations are basic for urban communities, utilities and different accomplices to streamline energy and asset stream to satisfy their grandiose zero-carbon objectives. The effective organization of local area energy frameworks dependent on disseminated sustainable power, stockpiling innovations and microgrids, for instance, requires investigation. Energy information is being utilized in a supportable energy in the executives framework (SEMS) that advances energy creation and utilization at the local area level, as indicated by projects like the EU-subsidized sharing urban communities programme (headed by London, Milan and Lisbon). The SEMS additionally associates with the bigger metropolitan information stage that is being made as a feature of the task.

Despite the fact that there is huge potential for more prominent information use across every single nearby help, city administrators should likewise conquer a few key hardships. A portion of these issues, like keeping up with information quality and distinguishing the vital destinations for each application, have, since a long time ago, tormented enormous scope information examination drives. Some particular issues accompany using large information and information investigation for city organization.

The smart city idea vows to incorporate information from a few sources, including various associations, distinctive environmental factors and a wide scope of clever devices. Indeed, even information incorporation inside associations is one of the most troublesome IT issues. Specialized impediments have been brought down by the reception of open principles all through the IT and correspondences enterprises, yet political and hierarchical hindrances are ordinarily the most difficult to survive. The current accentuation on making guidelines to direct keen city improvement is a critical stage towards tending to these difficulties.

The ascent of brilliant urban communities has introduced a huge number of new information streams with tremendous guarantee for working on neighbourhood administrations. This likewise puts urban areas at the bleeding edge of conversations over information possession and use. Many savvy city innovations will be acknowledged whether public and global information security guidelines, for example, the EU general information assurance guideline, are set up. A few urban communities, then again, accept they need to accomplish more. To meet the potential issues presented by the ascent of savvy urban areas, the association between occupants, government and specialist organizations should adjust. As large information, AI, and man-made reasoning become more normal in city organization and public and private administrations, this will just turn out to be more obvious.

Keen urban areas are entering an essential new period of advancement to accomplish huge upgrades in key markers and focused on results. This development requires the joining of advanced development and information-driven perspectives into administration plan and city arranging strategies. Information is used to improve navigation, empower continuous functional control, help administration quality and proficiency, and advance connection with individuals, organizations and different partners, which is the way the worth of savvy city innovation speculation has figured it out.

8.6.3 Data as a Service Drives Digital Transformation for Smart Cities

In the event that you drill down into a shrewd city's foundation, it doesn't take long to strike information. The foundation I'm discussing is one that joins data and correspondence innovation (ICT) into the texture of the city. The results are different savvy applications that make a superior resident encounter. These applications, driven by information, further develop regions like security, availability, versatility, actual framework, schooling, medical services, administration and supportability of the city.

The 'how' and 'why' of information-driven application stages.

One fruitful advancement approach is to carry out a flat and interoperable application stage layer that depends on an open and deft norm.

8.6.4 The Benefits of a Data-Driven Platform

1. Addressing citizen experience by delivering digital services to foster healthier living and better working environments
2. Incorporating local regulations with built-in security to protect user data
3. Delivering a continuous innovation platform that is extensible and flexible
4. Matching user demands with a responsive, agile platform
5. Developing a standardized IoT framework that ingests, analyses and presents myriad data points
6. Focusing on value creation rather than limiting to IT operations or systems integration

8.6.5 Data through the Hierarchical Layers of a Smart City

8.6.5.1 Sensor Layer

The hierarchy starts at the bottom with sensors as the basis of data generation. Sensors are devices attached to physical assets that generate data about that asset. Sensors can be specific to an application like parking, lighting, weather monitoring systems or incorporating multiple applications within one sensor like video cameras do.

8.6.5.2 Communication Layer

Sensors generate data that are passed through a communication layer. The communication layer can be wired and/or wireless depending upon the bandwidth of the data being transferred. In a complex environment, we notice devices with multiple types of communication protocols like NB-IoT, LoRa, CAT M1, LTE and others. The communication protocol needs to be able to multiplex various protocols to transfer data into our next layer, which is the digital platform layer.

8.6.5.3 Digital Platform Layer

The digital layer is diagnostic of the communication protocol that aggregates data from the various sensors. The aggregated data are either analysed on an edge server or placed in a cloud environment. Edge computing is primarily used for high data bandwidth applications such as videos, including analytics at the edge server. The other option, a cloud environment, can

- Intelligently monitor the collected data
- Analyse the data

- Implement automated predictive and prescriptive processes
- Digitalize and transform the data to develop capabilities, services and business models
- The data can be further analysed using AI and machine learning (ML) through anomaly detection and trend analysis

8.6.5.4 Data Layer

Lastly, the data layer incorporates other functions such as dashboarding, automated workflows and connecting to third-party applications and other city departments. Woven into this smart-city fabric is an additional element of IoT security to predict and prevent malicious activities.

8.6.6 Progressing to Data as a Service

A city's digital transformation strategy needs to collect and analyse data from various sources to continuously improve the efficacy of its applications. As cities go about weaving their layers, they're increasingly challenged by a lack of interoperability of the varied technologies, communication transmission protocols and architectures developed by various vendors.

8.6.6.1 Improving the Citizen Experience

Incorporating ICT into the city's fabric ultimately creates a better citizen experience. The key is the interconnectedness of the various assets driving data into decision-making and assisting with app development to improve livability.

8.7 CONCLUSION

As outlined throughout the chapter, companies are faced with substantial challenges in the context of digital transformation. In order to stay ahead of the competition, they are forced to innovate continuously with a strong focus on disruptive solutions creating real customer value. In this context, an increase in revenues may offset the tendency towards tighter profit margin. For companies addressing challenges in the field of smart city solutions, the awareness of the level of complexity and the multitude of facets is important. Thus, it will determine the strategic options and shape the innovation management suitable to any given organization. Going forward, additional research will be required to put the desk study results of this chapter to a practical test. In addition to that, smart city solutions are expected to evolve in the years to come, providing ample opportunities for further scientific groundwork both on a macro and on a company level.

Organizations need to improve to prevail after some time. A quickly developing piece of advancement comes straightforwardly from the most recent digitalization instruments. Numerous assets and a decent arrangement of considerations are coordinated to the further advancement of strategies, processes, devices and so on around digitization. New items and administrations in any remaining ventures, activity stream, modern offices, work association and, obviously, plans of action have been significantly affected by advanced innovations. To remain effective, new work abilities are required.

The impact and change coming from the digitalization falter are upsetting pretty much every industry in all cycles beginning with advancement underway in maintaining the business just as the board. The effect of digitization is enormous and is found in all ventures, achieving an interruption of existing plans of action through new innovation. Subsequently, organizations today need to open up their brains to advancement and constant learning. In the cutting-edge world, clients are better educated and are clear with regards to their assumptions. There is a positive shift towards individualized items and administrations just as an extraordinary client experience.

Digitalization, nonetheless, faces large difficulties. A machine or a sensor can't perceive a feline initially, for example. It needs many marked pictures. Building trust, which may be communicated by guaranteeing power, reasonableness, logic, ancestry of the information and results, needs still a ton of exertion.

The advanced change and the fourth modern transformation interlock different advances, like biotechnologies, nanotechnologies and quantum PCs (Schwab, 2016). Digitalization consolidates and even unites like never before all parts of an advancement framework, involving colleges and examination associations, industry, government, common society and media, or even the climate. The expansive foundational change calls for improved and joint activity in the development framework, in the financial framework, in the public eye, in political activity. The methodology of the Quintuple Helix may uphold the advancement of the overall government assistance of society and, subsequently, of the economy. Carayannis and Campbell have worked out the Quintuple Helix approach, among others.

Outlook changes inside businesses require a new socio-specialized framework to be made, with an alternate supporting foundation. Specialties moulding new mechanical standards might be relied upon to travel through specific stages. These variables comprise a succession of S-bends or waves, with the verifiable industrialization as the First Deep Transition (see Kanger & Schot, 2019). All development frameworks and all economy frameworks worldwide need to adapt to this change.

A lot of exploration is going on with the formation of an information base for an exhaustive comprehension of how digitalization manages the entire development framework, with industry, with the entire financial framework and with the entire society. Also, much examination is additionally expected to comprehend and to foster models for making it unmistakable in industry, government and society.

REFERENCES

Aliee, F. S., Kashfi, H., & Farahani, B., (2019). The evolving enterprise architecture: A digital transformation perspective. https://www.researchgate.net/publication/332281901_The_Evolving_Enterprise_Architecture_A_Digital_Transformation_Perspective

Anthony Jnr, B., (2019). A case-based reasoning recommender system for sustainable smart city development. https://link.springer.com/article/10.1007/s00146-020-00984-2

Anthony Jnr, B., (2020). Smart city data architecture for energy prosumption in municipalities: concepts, requirements, and future directions, *International Journal of Green Energy*, *17*, 827–845. https://doi.org/10.1080/15435075.2020.1791878

Anthony Jnr, B. (2021). Managing digital transformation of smart cities through enterprise architecture–a review and research agenda. Enterprise Information Systems, 15(3), 299–331.

Anthony Jnr, B., Majid, M. A., & Romli, A., (2018). An analytical study evaluating the applicability of a developed innovative e-sourcing system for automobile based firm. https://www.researchgate.net/publication/336255087_An_Analytical_Study_Evaluating_the_Applicability_of_a_Developed_Innovative_E-Sourcing_System_for_Automobile_Based_Firm

Babar, Z., & Yu, E., (2015). Digital transformation – implications for enterprise modeling and analysis. https://ieeexplore.ieee.org/abstract/document/8907284

Bokolo and Petersen (2019). A case-based reasoning recommender system for sustainable smart city development, DOI:10.1007/s00146-020-00984-2

Bibri, S. E. & Krogstie J., (2021), A Novel Model for Data-Driven Smart Sustainable Cities of the Future: A Strategic Roadmap to Transformational Change in the Era of Big Data. *Future Cities and Environment*, *7*, 1–25. https://ntnuopen.ntnu.no/ntnu-xmlui/bitstream/handle/11250/3043547/Bibri.pdf?sequence=1

Crawford, K., & Schultz, J., (2014). Big data and due process: Toward a framework to redress predictive privacy harm, Boston College Law Review, *55*(93), pp. 93–128.

Goerzig, D., & Bauernhansl, T., (2018). Enterprise architectures for the digital transformation in small and medium-sized enterprises. https://www.semanticscholar.org/paper/Enterprise-Architectures-for-the-Digital-in-Small-Goerzig-Bauernhansl/6a622d8cd158badb234e19b5a7ae7f711cb3f918

Gupta, D., Ghersi, F., Vishwanathan, S. S., & Garg, A., (2019). Macroeconomic assessment of India's development and mitigation pathways. https://doi.org/10.1080/14693062.2019.1648235

Hänninen, M., Smedlund, A., & Mitronen, L., (2017). Digitalization in retailing: Multi-sided platforms as drivers of industry transformation, *Baltic Journal of Management*, *13*(10). https://www.emerald.com/insight/content/doi/10.1108/BJM-04-2017-0109/full/html

Hausberg, J. P., Liere-Netheler, K., Packmohr, S., Pakura, S., & Vogelsang K., (2019). Research streams on digital transformation from a holistic business perspective: A systematic literature review and citation network analysis, *Journal of Business Economics*, *89*, 931–963. https://link.springer.com/article/10.1007/s11573-019-00956-z

Hermes, S, Riasanow, T., Clemons, E. K., Böhm M., & Krcmar H., (2020). The digital transformation of the healthcare industry: Exploring the rise of emerging platform ecosystems and their influence on the role of patients, *Business Research*, *13*, 1033–1069. https://link.springer.com/article/10.1007/s40685-020-00125-x

Ismagilova, E., Hughes, L., Dwivedi, Y. K., & Raman, K. R., (2019). Smart cities: Advances in research—An information systems perspective. http://hdl.handle.net/10454/16802

Kanger, L., & Schot, J., (2019) Deep transitions: Theorizing the long-term patterns of socio-technical change, *Environmental Innovation and Societal Transitions*, *32*, 7–19. https://doi.org/10.1016/j.eist.2018.07.006

Kitchin (2014, 2016). Big Data, new epistemologies and paradigm shifts. https://doi.org/10.1177/2053951714528481

Markovitch, S., & Willmott, P. (2014). Accelerating the digitization of business processes. McKinsey – Corporate Finance Business Practise, 1–4.

Mendhurwar, S, & Mishra, R., (2019). Integration of social and IoT technologies: Architectural framework for digital transformation and cyber security challenges. https://www.tandfonline.com/doi/abs/10.1080/17517575.2019.1600041

Nandram S. S., & Bindlish, P. K. (2017). Managing VUCA through Integrative Self-Management, Berlin: Springer. https://link.springer.com/book/10.1007/978-3-319-52231-9

Petrova-Antonova, D., & Ilieva, S., (2018). Smart cities evaluation – A survey of performance and sustainability indicators. https://www.semanticscholar.org/paper/Smart-Cities-Evaluation-%E2%80%93-A-Survey-of-Performance-Petrova-Antonova-Ilieva/d618a26977c734743fad2e45b92589f548d23f52

Pucihar, A., Ravesteijn, M. K. B. P., Seitz, J., & Bons, R. (2018, June). Managing complexity of digital transformation with enterprise architecture. In 31ST Bled eConference Digital Transformation: Meeting the Challenges (p. 635).

Ross, J., (2017). Don't confuse digital with digitization. MIT Sloan Review – September 2017; Retrieved 11 November 2018 from: https://sloanreview.mit.edu/article/dont-confuse-digital-with-digitization/

Schallmo, D., Rusnjak, A., Anzengruber, J., Werani, T., & Jünger, M., (2016). Digitale Transformation von Geschäftsmodellen: Grundlagen, Instrumente und Best Practices (Schwerpunkt Business Model Innovation), Wiesbaden: Springer Gabler, https://www.amazon.de/Digitale-Transformation-von-Gesch%C3%A4ftsmodellen-Instrumente/dp/3658123877/ref=asap_bc?ie=UTF8

Schwab, K, (2016). The fourth industrial revolution, *Quality Management Journal*, *25*, 108–109. https://doi.org/10.1080/10686967.2018.1436355

Shaikh, A. A., Sharma, R., & Karjaluoto, H. (2020). Digital innovation & enterprise in the sharing economy: An action research agenda. Digital Business, 1(1), 100002.

Skog, D. A., Wimelius, H., & Sandberg, J., (2018). Digital disruption, *Business & Information Systems Engineering*, *60*, 431–437. https://link.springer.com/article/10.1007/s12599-018-0550-4

Tomičić-Pupek, K., Pihir, I., & Furjan, M. T., (2019). Digital transformation playground, *Journal of Information and Organizational Sciences*, *43*. https://jios.foi.hr/index.php/jios/article/view/1302

Vial, G., (2019). Managing complexity of digital transformation with enterprise architectur understanding digital transformation: A review and a research agenda. https://www.researchgate.net/publication/331020687_Understanding_digital_transformation_A_review_and_a_research_agenda

9 Analysing the Application of a Smart City in Big Data Using K-Means Clustering Algorithm

V. Kavitha[1], *B. Sundaravadivazhagan*[2], *R. Karthikeyan*[3], *and P. Hemashree*[4]
[1]Sri Ramakrishna College of Arts & Science, Udayampalayam, India
[2]University of Technology and Applied Science, Al Mussanah, Oman
[3]Vardhaman College of Engineering, Shamshabad, India
[4]Coimbatore Institute of technology, Udayampalayam, India

9.1 INTRODUCTION

Big Data is a set of information that is large in quantity, developing greatly with time. It is information with a very massive volume and complex data where none of the conventional information control's equipment can keep it or operate it efficiently. Big Data refers to large-scale data management and analysis generation that exceeds the capability of traditional data processing generation.

Governments and municipalities adopt smart city (SC) initiatives to mitigate the difficult effects of nonstop urbanization trends and growing population density in cities. Generally, SC initiatives are executed through supplying residents at a higher class of existence in economically and environmentally sustainable cities. The idea of an SC is a huge factor for city development.

The goal of Big Data analytics is to implement specific safeguards in order to successfully gather high-brow information in real time. Even though Big Data analytics usually have a full-size assurance, there are a number of disturbing conditions that have to be resolved to realize its actual capability.

However, the connection among the underlying city domain names is an important idea to comprehend city smartness. This incorporated view of an SC implies cross-area sharing of information. From this perspective, an SC is considered an entire frame of incorporated structures or a machine of structures.

The promise of an SC ends in an exponential boom in facts through numerous orders of magnitude. Consequently, such vast volumes or massive facts are at the centre of the offerings rendered through the IoT. The phenomenon of Big Data has

DOI: 10.1201/9781003225317-9

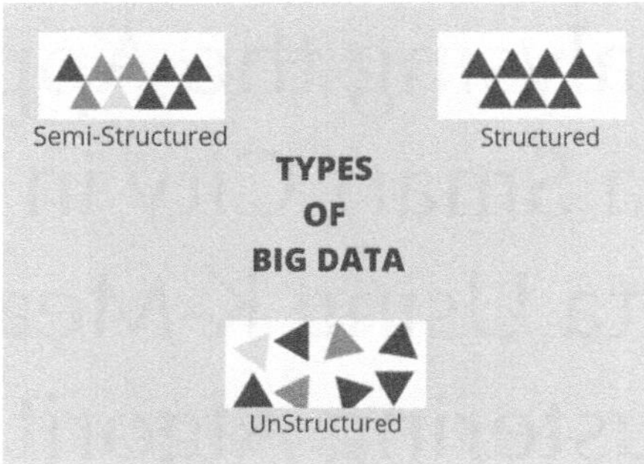

FIGURE 9.1 Big Data – types.

been characterised by volume, velocity and lots of acts sorts which have been created at ever-growing rates (Pramanik et al., 2017).

Figure 9.1 discusses about the types of Big Data. Generally, there are three kinds of Big Data structures namely structured, unstructured and semi-structured.

Figure 9.2 discusses about the applications of healthcare, transportation, power-grid, smart education, and energy. The volume of Big Data plays a vital role in the above-mentioned real-time applications.

9.1.1 Applications of Big Data

In the modern era, huge amounts of data are collected. Many industries take this data, turn it into Big Data and convert it into easy-to-use choices. Big Data have

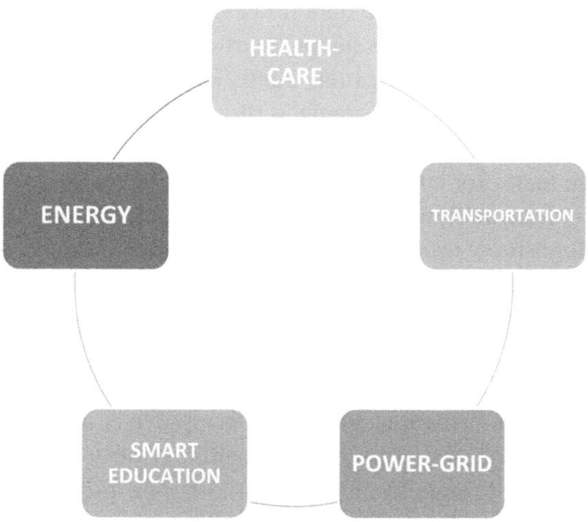

FIGURE 9.2 Big data applications in smart cities.

changed the perspective of view in today's world. Applications of Big Data are listed below (Townsend, 2013):

- Health sectors
- Education sectors
- Government sectors
- IoT
- Entertainment and media sectors
- Diagnosis and treatment
- Smart traffic system
- Automated driver in cars/automobiles

9.1.2 Role of Big Data in Smart Cities

SC packages generate big quantities of period even as large statistics structures make use of this statistics to offer facts to beautify SC packages. The large statistics structures will collect, progression, and search SC packages facts in a well-mannered way to provide facts to enhance one-of-a-kind smart metropolis services (Mohamed & Al-Jaroodi, 2014).

9.1.3 Benefits of Smart Cities

Presently, numerous cities participate to be clever cities in expectation of gaining cautiously, ecologically, and publicly. As a result, many also are eyeing the potentials made considerable through method of means of the procedure of applications of SC analytics in Big Data. Hence, this stage has a number of blessings and opportunities which could contribute in manufacturing the collection to alter or redecorate an urban to an SC. With such selection, it is able to be viable to gain better layers of sustainable, flexible, and authorized (Su et al., 2011).

In addition to enhancing the public's first-class of presence and presenting smart mechanism of organizations and assets.

Profits of smart cities include the following:

- Effective resource consumption
- Improved quality of life
- Advanced levels of transparency and openness (Chourabi et al., 2012).

9.1.4 Big Data Applications Serving Smart Cities

- **Smart education:** The education developments are adeptness, usefulness, and productiveness. The use of education offerings that might be flexible and smart to offer a higher use of information better help for life getting to know for all human beings. Smart education packages cooperate human beings in active receiving to recognize surroundings that allow them to follow quick adjustments of humanity and the atmosphere.

- **Smart traffic lights:** The key factors of smart city are to manage the traffic lights in the city, to develop the traffic arrangements, and recover the inhabitants' commutes and the city's distinctive traffic patterns. While the people rise, traffic stresses, fumes, and financial troubles arise (Kumar & Prakash, 2016).

Other Big Data applications serving smart cities include the following:

- Smart network
- Smart hospital
- Smart dynamism
- Smart vehicles
- Smart surroundings
- Smart security
- Smart domination

9.1.5 Why Implement Big Data in Smart Cities

In the modern era, Big Data is a crucial feature improving the world. One of the improved technologies in smart cities is using Big Data analytics. The key tenacity of introducing Big Data in an SC is to decrease traffic accidents and congestion in the city. Big Data also helps in managing data storage in smart cities. Resources and services can also be improved in smart cities with the help of Big Data (Sivakkolundu & Kavitha, 2019).

9.2 LITERATURE SURVEY

The combination of Big Data analytics along with SC has amplified in modern years, revealing the attachment among the two broad areas. Governments and municipalities of every country undertake the projects of SC to moderate the demanding impacts of permanent urbanization enhancements and growing population concentration in urban areas.

In general, SC projects intend to propose improved environmental quality of life for humans as well as cost-effectively sustainable cities. For urban modernization, the concept of an SC is a very recent one. Even though the expression is utilized expensively in the literature of academic as well as urban enhancement projects, there is no common consensus regarding the term "smart" in the background of smart cities. Spontaneously, the SC consists of elegant elements. The conflict is not worthy, as smart elements do not represent an SC, but sustaining the city's elements' interrelationship is a necessary trait for a city to be smart. From the perspective of information systems, the specific feature refers to gathering, analysing as well as exchanging information among numerous city domains.

The SC is a well performing city composed of smart fields. These are smart living, smart governance, smart mankind, smart surroundings, and smart mobility. Although researchers disagree on the number and character of smart domains that comprise a SC, there is universal agreement that a religion's smartness is recognized by maintaining cross domain sharing and interplay of statistics. Data-driven decision-making is a current trend of a Big Data analytics major enabler that is increasing interest in information security, which signifies the practice of basing judgements on information investigations rather than solely on intuition.. Numerous techniques are utilized in Big Data analytics, comprising machine learning, artificial intelligence deep learning as well as neural networks. Traditional statistical techniques of data analysis are to be compared with these modern technologies that are handling unstructured as well as structured information and extracting qualitative and quantitative indicators.

At present, over 50% of the globe's society live in cities, and this statistics will exceed 70% by 2050. Urbanization will speed up in the subsequent 30 years. Generally, cities occupy 1% of land area, consume 75% of energy as well as 85% of resources and also produce 80% of greenhouse gas emissions around the globe. With the endless urbanization, cities will handle more sensitive environmental pressure as well as contradiction among demand and supply of resources. On the other hand, speedy urbanization has also caused a succession of urban problems like traffic congestion, contradiction among demand and supply of resources, and environmental pollution. The hurdles and obstacles in the process of urbanization have become one of the most vital hurdles facing sustainable growth of all countries in the world.

9.3 CHALLENGES OF BIG DATA IN SMART CITIES

Every field has its own challenges and limitations. This field's challenges include the following:

- SC population: Smart population is one of the difficulties in which excessive population results in rapid expansion of an SC, such as traffic pollution, city congestion, and social equality. The population of a city has a significant impact on the magnitude of Big Data.
- **Cost:** Cost is another key challenge. SC growth always includes a budget. It requires a large amount of money. Infrastructure is the key concept to be involved in smart cities. Without enough money, nothing can be built. Funding is be a greater challenge in an SC.
- **Lack of skilled professionals:** Without proper technicians, the project won't look good and effective. To prepare a proper strategy to build, experienced experts are required. Every field requires experienced workers. Skill is most important in the field. Planning and undertaking the scheduled strategy accurately is not an easy task.
- **Security risks:** In order to protect an SC, proper security maintenance must be done. By constructing effective projects without not securing them, everything is in vain. Basic security will be implementing CCTV in every corner of the cities.
- **Data and sharing information:** Gathering data from civilians and distributing it to other citizens without their knowledge is a major challenge. Each of the government departments has the confidentiality of public's information (Pourzolfaghar et al., 2018).

9.3.1 Crucial Challenge: Security Risks

The key challenge is security risks. In smart cities, databases are confidential to the government and some private sectors. For this, a smart level of safety is required. For every SC without proper privacy and security the public and many sectors in the city will be in danger. Making certain the confidentiality of public's protection is a major task.

In order to rectify this challenge, governments should properly plan with experienced experts in the field. Introduce CCTV cameras in every corner of cities. Proper anti-virus software must be installed to secure the data. Update it to the newest software. Backup should be kept accordingly as it holds the entire city's confidential information (Kavitha et al., 2020).

9.3.2 Necessities for a Smart City

For an SC to develop, proper requirements should be available. These include the following:

- Big Data treating stages
- Smart system organization
- Innovative procedures
- Typical skills
- Safety and confidentiality
- Resident consciousness
- Administration part

9.4 METHODOLOGY

The tool used here is the Orange 3.31.1 version. It is for predicting data analysis. Orange is a data visualization toolkit used for interactive information visualization and explorative data evaluation. The Orange tool is selected because it is user friendly and comprises an easy approach to finding the data analysis. A sample pattern screenshot is given in Figure 9.3, which shows the workflow of the analysis.

There are more methods and algorithms accessible in the Orange tool, such as SVM, Ada booster, linear regression, PCA, Louvain algorithm, K-means, kNN, Radviz, Random Forest, Heat Map, etc.

The dataset is a CSV file downloaded from the Kaggle.com; primarily, the CSV file import icon is established to load the dataset from the CSV file. In the next stage, the clustering algorithm called the K-means clustering algorithm is applied. The next phase is to join the scatter plot to the clustering algorithm and see the plotted graph. This designed graph will be viewed in a separate window called a scatter plot. In this window, the x-axis and the y-axis can be selected, a combo box is delivered for that use, and also the sorting can be done using varieties of colours.

K-means clustering algorithm is used here because this algorithm easily adjusts to new models. The collected data can be analysed using different colours and sizes. No matter how big the variables and data, a K-means clustering algorithm responds

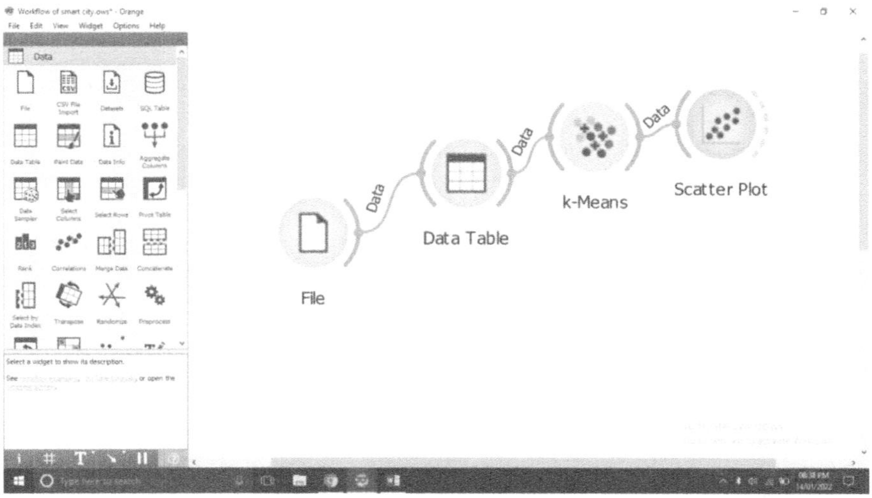

FIGURE 9.3 Screenshot of workflow in the Orange tool.

faster than other clustering algorithms. That's why K-means is ideal in the Big Data in SC dataset (Kavitha et al., 2019).

9.4.1 K-Means Clustering Algorithm

With the increasing use of the Internet in today's civilization, the amount of statistics produced is enormous. Even though the nature of human being data is undemanding, the absolute amount of information to be investigated makes processing complicated for even computers. To handle such processes, the machine learning techniques of clustering enable us to inspect huge quantity of information in an intelligible manner.

K-means is one kind of clustering algorithm, which belongs to unsupervised learning process that is used to solve the grouping obstacles with the assistance of the recent techniques, such as data science, artificial intelligence, machine learning, deep learning, neural networks, etc. This algorithm is used to group the unlabelled dataset into numerous expected clusters. This clustering technique acquires the unlabeled dataset as input and divides the datasets into n number of clusters and repeats this task The value of n is predetermined. Through this algorithm, the count of predefined clusters is already defined before executing the grouping process.

For instances, if the number of clusters is defined as two, two clusters are generated, whereas if the number of clusters is defined as three, three clusters are framed; likewise a certain number of clusters are generated according to the requirement and so on. The K-means technique allows us to organize the provided information into several groups in a comfortable manner to discover the various clusters in the unlabelled dataset on its own, eliminating the need for multiple training sessions. K-means clustering is a centroid-based technique, where every cluster is connected with a centroid. The main intension of this technique is to reduce the summation of distances among the dataset objects and their subsequent clusters.

The K-means clustering algorithm essentially achieves the following two missions:

- Determines the most excellent value for K centroids by an iterative progression.
- Allocates each data point to its neighbouring k-centre. Those data points which are close to the exacting k-centre produce a cluster.

The working structure of the K-means clustering algorithm is elucidated as follows:

Step 1: Give the value for the predefined number of clusters K.
Step 2: Choose K centroids randomly.
Step 3: Allot every dataset point to their nearest centroid point, which will outline the predefined K number of clusters.
Step 4: Estimate the conflict and position a new centroid of every cluster.
Step 5: Reiterate the third step, which means relocate every dataset point to the new neighbouring centroid of every cluster.
Step 6: Whether any relocation occurs, it goes to the fourth step or else goes to finish.
Step 7: Stop the model process.

The basic K-means procedure can be described as follows:

Figure 9.4 portrays the working principle of the K-means clustering technique. This clustering technique is imitated with a predefined number of clusters K and

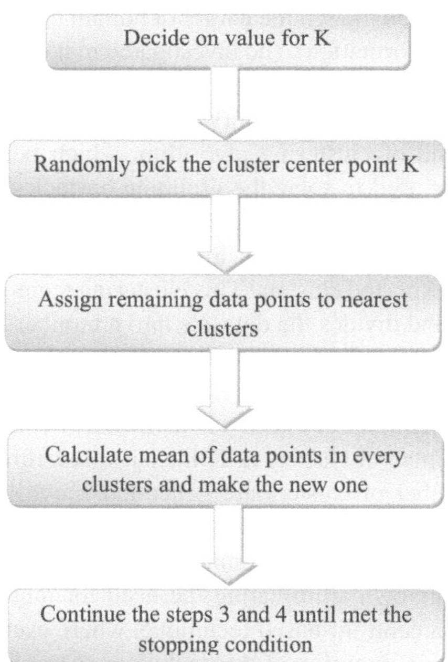

FIGURE 9.4 Working principle of K-means clustering.

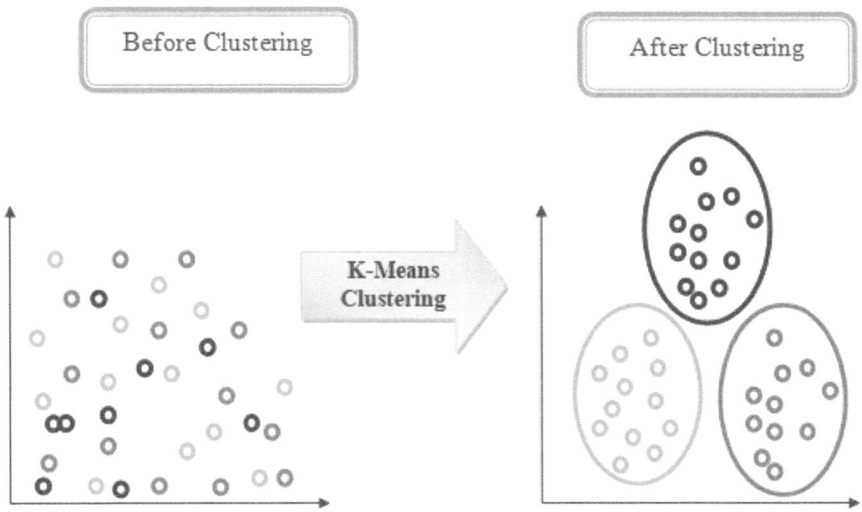

FIGURE 9.5 Working structure of K-means clustering technique.

according to this value, the clustering process starts. Generally, the initial step itself is to be determined that the required number of clusters is fixed. The cluster centroid points are randomly picked for each and every cluster. Based on that centroid point value, the remaining cluster point's distance is calculated with their similarity values. The remaining data points are measured with their similarity, then the nearest values are determined with the individual clusters. Simultaneously, the mean value of each cluster data point is calculated and a new group is formed. Continuously, this process is continued until the stopping condition is met. Like this, the clusters are discovered and formed using the clustering technique of K-means clustering.

Figure 9.5 of the K-means clustering technique explained the working procedure of the grouping process with various data object values. Before the clustering process, every kind of data object is assembled in a particular platform. After applying the K-means clustering procedure, the data objects are formed in a particular group according to their similarity. Hence, three kinds of data objects are mixed up with the whole dataset that will be divided into three cluster groups according to their similarity values.

9.4.2 Various Kinds of Distance Measures in K-Means Clustering

For K-means clustering, the distance among each and every data point of the dataset needs to be calculated, and all centroid data points are initialized. Belonging to the discovered values, any specific data point is assigned as the centroid point with minimum distance. Consequently, the estimation of distance plays a vital role in the clustering technique. Hence, calculating the distance among the two data points can be estimated with various kinds of distance measure techniques. The distance matrices of Euclidean distance measure, cosine distance measure, and city block distance measures are considered for calculating the distance value among the various data points in the K-means clustering algorithm.

Euclidean Distance Measure

- It works great when the dimensionality of the data is low.
- For obvious reasons, it is usually used to measure distance.
- Extremely perceptive to use.
- Simple to employ.
- Shows great outcomes in most of the use cases.
- Very simple to test.
- Will not support when the data is image.
- The largest value of the variables will significantly influence the result.
- Does not operate creatively with picture data.

Cosine Similarity Measure

- It works great when the dimensionality of the data is high.
- It is very simple to measure the cosine of the angles among two vectors.
- Mutually continuous and categorical variables may be used.
- Doesn't work economically with nominal data.

City Block Distance Measure

- Simple to generalize into higher dimensions.
- Doesn't work resourcefully with image data.
- Could not be used to categorize documents.
- Imagine vectors that portray objects on a consistent grid such as a chessboard.
- City block distance measure is also termed Manhattan distance measure.
- Manhattan distance measure is less sensitive than Euclidean distance while being employed in high-dimensional data.
- The shortest path is not possible to supply.
- It gave elevated distance value than the distance metric of Euclidean as it doesn't possible to provide a shortest path.

Comparing the above-mentioned distance measures refer Figure 9.6, the Manhattan distance is considered an optimal measure for calculating the distance value among the desired two data points. Hence, the Manhattan distance measure is also known as the city block distance measure. This measure is very much suitable for the machine learning technique of the k-means clustering algorithm because of its high dimensionality. The concept of the Manhattan distance measure is initiated with selecting any two data elements around the entire dataset.

The distance measure of the Manhattan distance is also known as in two forms. The former one is the city block distance measure and the later one is the taxicab distance measure. It aims at measuring the distance among the vectors of real valued. These vectors describe the objects on a standardized grid like a chessboard. It means that the distance computation will only proceed at right angles.

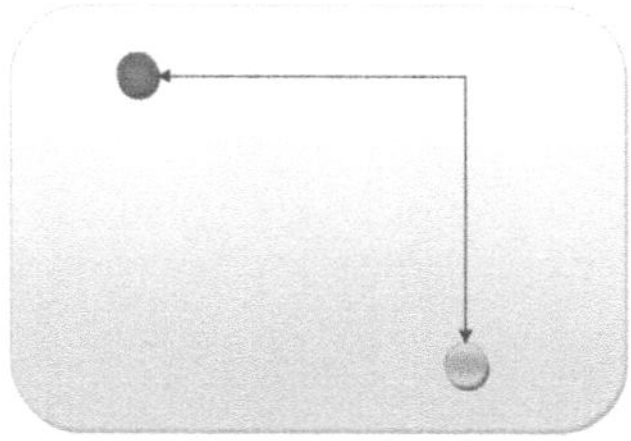

FIGURE 9.6 Distance value calculation using the Manhattan distance measure.

$$Dis(x, y) = \sum_{i=1}^{k} |x_i - y_i|$$

where D is is the variable to calculate the distance among the data objects x and y, and a k number of iterations are to be considered to form the optimal clusters. The nature of the Manhattan distance measure is very simple to form the best clusters, which is more susceptible than the conventional distance measure of the Euclidean distance measure while working with high dimensional dataset. Moreover, it could not provide the shortest path. Simply stated, the Manhattan distance measure is very much suitable for calculating the finest distance value for predicting the valuable clusters with the assistance of the k-means clustering algorithm. Hence, the Manhattan distance measure is applied to the k-means clustering algorithm for calculating the distance value. It signifies that the distance will only be computed at straight angles.

9.5 IMPLEMENTATION

The dataset used in this study is an acorn_details.csv file that consists of a smart meter in London. The dataset is occupied from kaggle.com. The K-means clustering algorithm is used in this dataset to demonstrate the data analysis.

In this dataset, the energy consumption analysis of 5567 models in UK power networks project from November 2011 to February 2014 is enclosed. The goal is to construct smart meters in almost every home, refer Fig 9.7 (Sathishkumar et al., 2018).

9.6 RESULTS AND DISCUSSIONS

The K-means clustering algorithm is the finest selection to produce results for executing the analysis. This sector includes the captured screenshots of the K-means clustering algorithm, demonstrating the data of energy consumption readings in five different smart cities.

In Figure 9.8, City A readings are depicted in which city readings went above 0.64% within 200 km. Three different smart cities are represented in C1, C2, and C3 with the colours blue, red, and green, respectively, shown in the graph. C2 shows lower % readings compared to others. This happens because C2 is set up in a distance of kilometres (3000 km). C3 went up to 0.6% within a radius of 100 km. This graph proves that C1 city is better and efficient compared to C2 and C3 (Alexandrov et al., 2014).

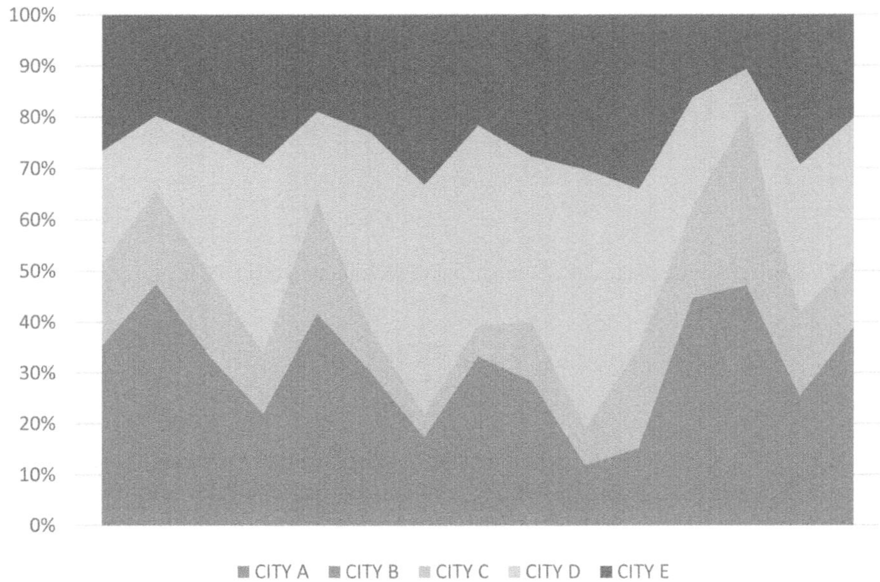

FIGURE 9.7 A sample graph with respect to Table 9.1 – total area covered in categories of energy consumption readings in five different smart cities.

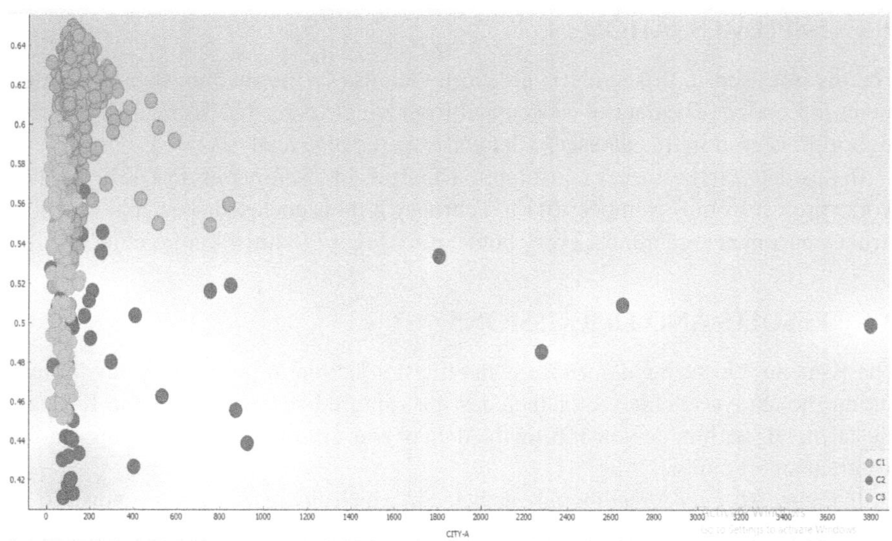

FIGURE 9.8 K-means clustering algorithm graph of energy consumption readings in which x-axis is for total kilometres covered and y-axis is for % of energy consumed in the smart meter.

TABLE 9.1 Housing Data
First 15 Rows of Dataset Taken Named "acorn.csv" File. The Dataset Contains Eight Distinct Columns such as Main Categories, Categories, References, and Five Different Cities in London

Main Categories	Categories	Reference	City A	City B	City C	City D	City E
Population	Age	Age 0–4	77	83	72	100	120
Population	Age	Age 5–17	117	109	87	69	94
Population	Age	Age 18–24	64	73	67	107	100
Population	Age	Age 25–34	52	63	62	197	151
Population	Ethnicity	White	98	106	110	83	93
Population	Ethnicity	Mixed	117	79	55	248	149
Population	Ethnicity	Black	48	34	21	215	158
Population	Ethnicity	Other ethnicity	174	84	47	304	168
Housing	House tenure	Shared/equity ownership	84	78	67	184	158
Housing	House size	Number of beds: 1	29	23	32	223	133
Housing	House size	Number of beds: 2	27	28	72	113	124
Economy	Economic activity	Self-employed	169	132	120	144	109
Economy	Economic activity	Retired	111	106	154	40	49
Economy	Economic activity	Unemployed and seeking work	29	42	46	81	82
Economy	Economic activity	Student	110	80	66	133	99

In Figure 9.9, the demonstration of three types of meters is shown. C1, C2, and C3 are three different smart cities characterized in blue, red, and green, respectively. C1 shows higher % readings which are above 0.64%. C2 is around 0.58% and C3 is around 0.6% energy consumption readings. Another result shown in Figure 9.6 is C2 city energy readings, the lowest point of which is below 0.42%. This graph proves that C1 is better than other two cities (Ashik et al., 2021).

In Figure 9.10, total kilometres covered to set up the metre for three various smart meters are shown. C1, C2, and C3 are the three varieties of smart cities represented in blue, red, and green colours, respectively. From the graph, it is clear that type C1 is covered the most, which is an almost 420 km radius. The lowest kilometres covered are C3, which is around 100 km radius. C3 covers up to 360 km. From this, C1 requires more funds as it covers an almost 420 km radius. The average kilometres covered would be a 100–140 km radius.

The previous figures mentioned that varieties of main categories involved in the smart cities are shown. Figure 9.11 shows main categories that include community safety, contact, digital, economy, education, and environment. C1, C2, and C3 are three altered smart cities represented in blue, red, and green, respectively. In C1, the digital sector is used more compared to C2 and C3. C1 is using more main categories, whereas C2 is using the least (Iwendi et al., 2020).

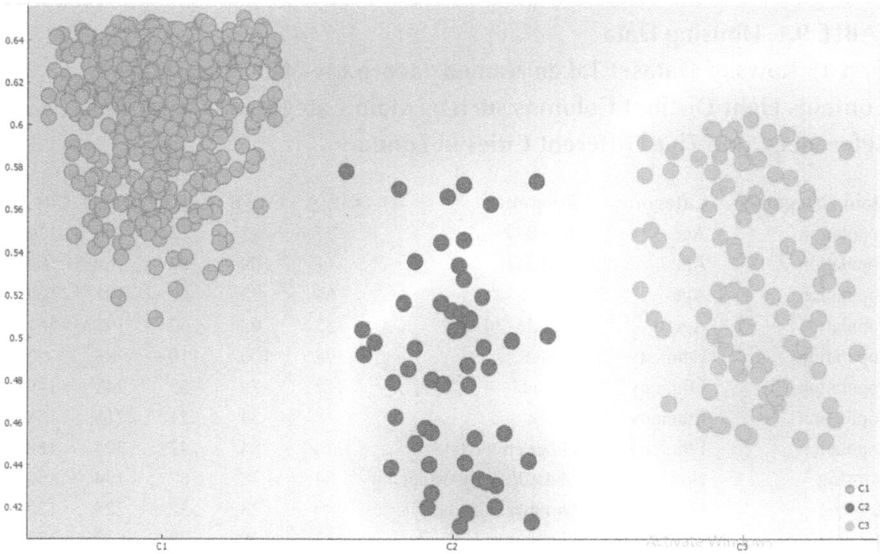

FIGURE 9.9 K-means clustering algorithm graph of energy consumption readings in which the X-axis is types of meters C1, C2, C3, and Y-axis is % of energy used in the smart meter.

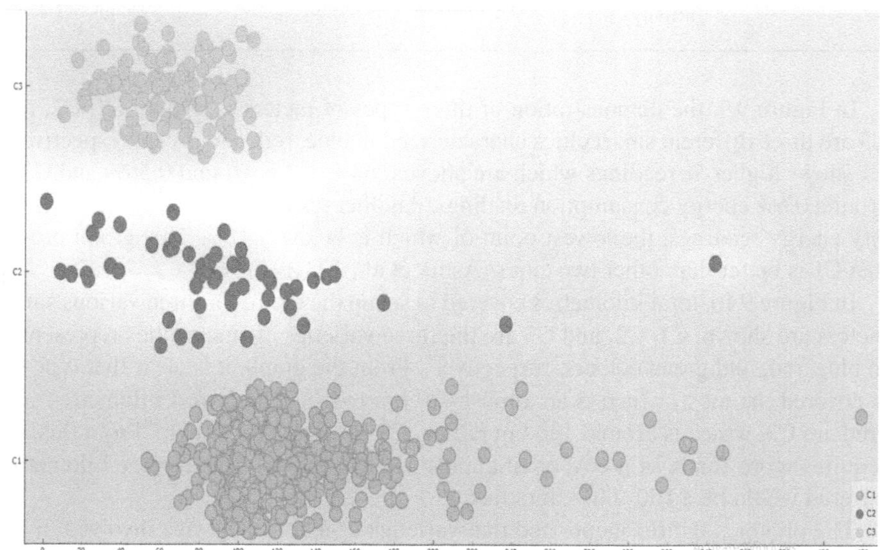

FIGURE 9.10 K-means clustering algorithm graph of energy consumption readings in which X-axis is for total kilometres covered and Y-axis is for three different smart cities, C1, C2, and C3.

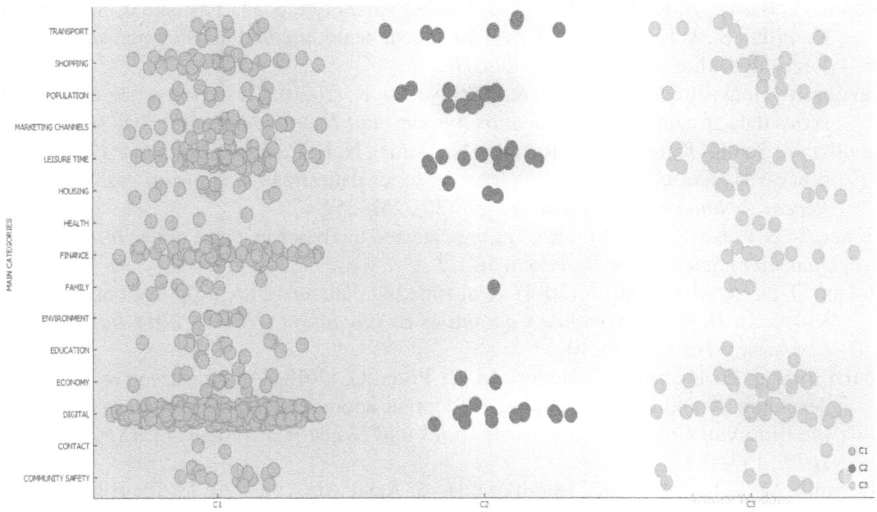

FIGURE 9.11 K-means clustering algorithm graph of energy consumption readings in which X-axis is types of meter C1, C2, and C3, and Y-axis is main categories/sectors in the smart city.

9.7 CONCLUSION

Big Data can help smart cities in an efficient way. A SC in addition to Big Data are two recent and significant concepts. In this study, applications of Big Data serving smart cities were discussed. Various challenges of Big Data in an SC were also discussed. The main challenge is security risks in the smart cities. Big Data requirements helping smart cities are a mutual phase in this study, which helps cities to advance in an effective way. In this chapter, the crucial challenge and security risk solutions were discussed. A demonstration of graphs using five varieties of smart cities was depicted. The cities that make use of current and latest technology will be developed and make the city smart. The main approach needed to make a smart city is proper planning. For that, professional technicians are required. This is how smarter city security using Big Data is improved.

REFERENCES

Alexandrov, A., Bergmann, R., Ewen, S., Freytag, J. C., Hueske, F., & Heise, A., et al. (2014). The stratosphere platform for big data analytics. *VLDB J*, *23*(6), 939–964.

Ashik, M., Subashini, B., & Sundaravadivazhagan, B. (2021). Amalgamation and characterization for test suite enhancement in software testing using data mining techniques. *Materials Today Proceedings.*

Chourabi, H., Nam, T., Walker, S., Gil-Garcia, J. R., Mellouli, S., Nahon, K., & Scholl, H. J. (2012). Understanding smart cities: An integrative framework. In *System Science (HICSS), 2012 45th Hawaii International Conference on IEEE*, 2289–2297.

Iwendi, C., Bashir, A. K., Peshkar, A., Sujatha, R., Chatterjee, J. M., Pasupuleti, S., Misshra, R., Pillai, S., & Jo, O. (2020). Covid-19 patient health prediction using boosted random forest algorithm. *Frontiers in Public Health.*

Kavitha, V., Hemashree, P., Dilip, H., & Elakkiyarasi, K. (2020). Clustering process with time series data stream. *Smart Innovation Systems and Technologies, 159*, 335–343.

Kavitha, V., Senthil Kumar, A. V., Revathy, N., Daniel, N. K. C., & Hemashree, P. (2019). Pre-processed hierarchical clustering for time series data streams. *International Journal of Recent Technology and Engineering, 8*(30), 251–254.

Kumar, S., & Prakash, A. (2016). Role of big data and analytics in smart cities. *International Journal of Science and Research, 5*(2), 7.

Mohamed, N., & Al-Jaroodi, J. (2014). Real-time big data analytics: Applications and challenges. In *High Performance Computing & Simulation (HPCS), 2014 International Conference IEEE*, 305, 310.

Pourzolfaghar, Z., Bremser, C., Helfert, M., & Piller, G. (2018). A comparison of smart city development and big data analytics adoption approaches. In *Proceedings of the 7th International Conference on Smart Cities and Green ICT Systems (SMARTGREENS 2018)*, 157–164.

Pramanik, M. I., Lau, R. Y. K., Demirkan, H., & Azad, M. A. K. (2017). Smart health: Big data enabled health paradigm within smart cities. *Expert Systems with Applications*, 370–383.

Sathishkumar, B. R., Sundaravadivazhagan, B., Martin, B., Sasi, G., Chandrasekar, M., Rakesh Kumar, S., Elamaran, V., Balaji, V. S., & Arunkumar, N. (2018). Revisiting computer networking protocols by wireless sniffing on brain signal/image portals. Neural Computing and Applications, *2018*, 1–13.

Sivakkolundu, R., & Kavitha, V. (2019). Tobit Regressive based Gaussian independence Bayes map reduce classifier on data warehouse for predictive analytics. *International Journal of Innovative Technology and Exploring Engineering*, 4269–4280.

Su, K., Li, J., & Fu, H. (2011). Smart city and the applications. In *Electronics, Communications and Control (ICECC), 2011 International Conference on IEEE*, 1028–1031.

Townsend, A. M. (2013). *Smart cities: Big data, civic hackers, and the quest for a new utopia.* WW Norton & Company.

10 Artificial Intelligence for Smart City Development in BRICS Economies

Opportunities and Challenges

Roli Raghuvanshi[1], *Tanushree Sharma*[2], *Ravinder Rena*[3], *and Rashmi Rai*[4]
[1]University of Delhi, New Delhi, India
[2]Manipal University Jaipur, Jaipur, India
[3]Durban University of Technology, Durban, South Africa
[4]Christ University, Bangalore, India

10.1 INTRODUCTION

Professor John McCarthy in 1956 gave his views that artificial intelligence (AI) delegates "the science and engineering of making machines intelligent, especially intelligent computer programs." Stanford University's panel-AI 100 designated AI as "that quality that enables an entity to function appropriately and with foresight in its environment." More definitions regarding AI reiterate that AI acts as a computerized system that acts and thinks like mankind. It is something that stimulates perception, spatial processing, reasoning, environmental scanning and much more. AI creates an environment where continuous scanning of the environment goes on, and experience is gained from the same and measures in lieu of what they feel or sense are taken. Depending upon the cognitive abilities that are simulated, we observe three categories of AI that are applied.

Basic AI drives cognitive potential such as language, memory and attention, along with some executive functions like expectation and decision-making. It enhances the performance of business analytics solutions and improves the functioning of digital platforms. Examples include online matching, chatbots, credit scoring and smart speakers. **Advanced AI** explores further in the simulation of human cognitive abilities such as perception, sight, and spatial processing. It imitates the human mind and delegates the anatomization of unstructured data such as images, audio data and texts. Instantaneous applications include recognition of speech, health diagnoses, transport, urban planning, logistics, safety and transportation. **Autonomous AI** is expected to be self-driven, with the capability to interconnect with mankind and acquire skill independently. Autonomous AI is neither still nor very widely applicable due to its certain limitations.

DOI: 10.1201/9781003225317-10

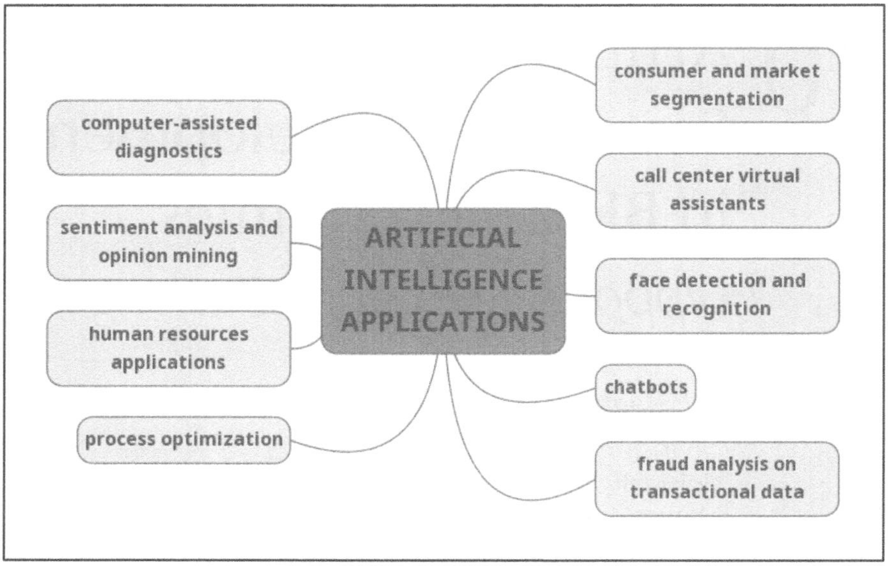

FIGURE 10.1 Applications of artificial intelligence.

1. Definition attributed to Professor John McCarthy in 1956. Quoted in Moor, James. 2006. "The Dartmouth College Artificial Intelligence Conference: The Next Fifty Years." *AI Magazine*, Volume 27, Number 4, pp. 87–91. https://pdfs.semanticscholar.org/d486/9863b5da0fa4ff5707fa972c6e1dc924 74f6.pdf.
2. See One Hundred Year Study on Artificial Intelligence (AI100). "Defining AI." Stanford University.

Today, when we look at the cities, we find them to be sitting on a plethora of data, which are produced through usage of Internet of Things (IoT), sensors, smart meters, telecommunication devices, traffic management systems, installed cameras etc. Already the archives for many years are available to the government and agencies; on the other hand, these smart devices are icing on the cake which is playing a very crucial role in city management. The urbanization process and migration of labourers from rural areas at a fast pace are making the application of AI inevitable for the economies, especially the emerging economies. We can certainly look forward towards an exponential growth using the available data and smart technologies. AI has the potential to make usage of the ample data and ameliorate the performance of cities, resulting into the optimization of operational resources and operational cost. The role played by Information and Communication Technologies (ICT) in smart city development widely covers these areas (Aguilera et al., 2017; Allam, 2017a; Neirotti et al., 2014). ICT technologies are being continuously developed and adopted in order to build platforms on which businesses, governments and citizens can collaborate and communicate, as well as to provide the connections between the networks (of people, businesses, technologies, infrastructures, energy and spaces) that serve as the foundation for the city's services. There is a tremendous amount of data of various forms and from diverse sources that may be shared and transferred due to the remarkable breadth and diversity of applications that are emerging in this setting (Brutti et al., 2019). Events that have changed many people's lives and had an impact on many different professional fields have occurred in the year 2020 (Mazzara et al., 2021). AI has helped the local bodies, decision-making entities and government regarding many important issues (Abaker et al., 2016; Souza et al., 2019). Smart cities have espoused the IoT as a mode to inspire orderliness and performance in urban affairs. AI paved the way for a platform that assists in gathering, analysing and distributing the data which plays a very prominent role in different spheres of life (Batty, 2016; Bhadani, 2016). In the case study of London, it was observed that AI was smart enough to capture the locations which had a problem of contaminated water, and proper measures were taken to avoid it happening any further to deal with the high mortality rate cause from the contaminated water (Barkham et al., 2018).

Big Data from AI also supports the attainment of economic goals with higher returns. Due to the implementation of AI in the process of developing smart cities, we see the availability of better supply chains, advanced infrastructure, good connectivity, real-time settlements, and high computing capabilities, making it possible to have a competitive edge over others (Davenport & Ronanki, 2018). The collaboration among mankind, machines and Big Data enables enhanced governance, and blockchain technology here plays a very significant role, contributing to the management of the cities (Voto, 2017). The generation of Big Data related to the stakeholders of the city will help the agencies and governments to reach out to the citizens to get their view on city governance, and this will contribute towards inclusive decision-making (Muzammal et al., 2019). The cities are witnessing vast transformations which in turn are gearing the demand for resources and land (Al-Kodmany, 2018). AI is creating vast opportunities for the humans as it is ensuring safety, quality education, and better job opportunities (Allam et al., 2018a; Allam et al., 2018b; Siew & Allam, 2017). The first Sustainable Development Goal (SDG) with a solely urban focus was established at the UN-HABITAT III conference in Quito in late 2016. Through a variety of measurements, indicators and evaluation systems, the so-called SDG 11 seeks to improve the inclusiveness, security, resiliency and sustainability of cities (Caprotti et al., 2017).

Figure 10.2 depicts the various advantages that come out of applications of AI in smart city development, and the cities witness innovative products and effective business models, including remarkable solutions to business, households and government, leading to tremendous results in the field of affordable cost, development of human resource and innovative services rendered by government agencies.

The AI is most suited to urban affairs as it enables personalized delivery of services, improves city planning, improves financial management of the city, improves coordination between the stakeholders, manages resource constraints, increases the utility of existing services and platforms, improves access to city infrastructure and foremost helps in planning for the economic uplift of the city (Raghuvanshi & Sharma, 2023). Direct and indirect benefits could be generated through the application of AI in the cities of emerging economies. The betterment of cities could be

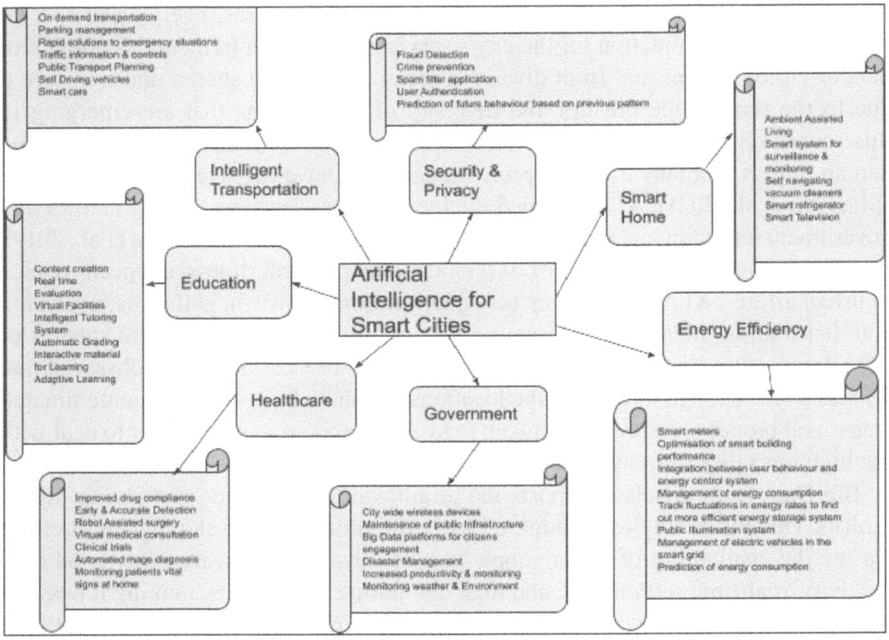

FIGURE 10.2 Artificial intelligence for smart cities.

experienced through quick, smooth and scalable solutions, and the citizens would experience a better lifestyle. AI is already making a significant beneficial difference in a wide range of fields related to the global economy and society. Humanitarian organizations, for instance, use intelligent chatbots to offer psychological assistance to Syrian refugees, and medical professionals use AI to create individualized cancer therapies (Castro & New, 2016). AI with citizen participation in urban areas affects how well green technologies work in contemporary cities, notably in terms of trash management. Therefore, the adoption of green technologies can only be made possible by a combination of the AI and the well-established behavioural attitudes of city dwellers (Mingaleva et al., 2020).

Though the implementation of AI in smart city development reaps a plethora of benefits, it is observed that AI poses certain threats also, which need to be addressed. For instance, challenges occur due to large-scale automation, like the problem of unemployment, which is bound to happen with the implementation of AI and machine learning (ML), daily increase of digital divide and technological divide, time lag in adaptation of new job requirements, privacy of stakeholders, trust issues and security matters. As we depend on Big Data, complexity cannot be overruled (Bari, 2017; Fan, Han, & Liu, 2014; Tole, 2013). Storage problems, spanning and scaling are big challenges of Big Data. As Big Data is related to big sample sizes, the problem of incidental endogeneity, accumulation of noise, factitious correlation and errors in measurements can be responsible for delayed results (Fan et al., 2014). Planning strategies for smart cities have gotten a lot of traction recently. While supporters of smart cities claim that the use of ICTs, improved governance and citizen capital would result in beneficial social transformation, detractors highlight the negative repercussions and flaws in their development and implementation. There has been much criticism of the idea and the way it has been embraced and implemented, despite efforts to encourage its usage (Kummitha & Crutzen, 2017). The ability to disseminate, investigate and gather information, using various devices in real-time adds to the problems, and the issues of privacy and psychological comfort is taken care of for urban people (Davey & Tatnall, 2015). Due to technological reasons, the possibility of accessing private data is a major cause of concern as it hinders the privacy of urban dwellers (Hildebrandt & Gutwirth, 2008). The social platforms make the private data public, thus posing a threat to one's privacy (Santanen, 2019; Zimet & Jacob, 2001).

These projects provide significant incremental gains for their target groups, according to analysis across procedural, rights, instrumental and structural aspects. However, those who benefit most are outside parties and wealthy areas, which raise relative inequality (Heeks & Shekhar, 2018).

The absence of homogeneity among various devices, collection of large data and proper analysis enormous; information and data security concerns, system failures, proper waste management, cost of operation, performance, connectivity issues and sustainability are some of the major provocations. The development of the smart city faces the challenge of maintenance and design. Managing operational as well as design cost is a big deal, especially with the emerging economies. To gain cost optimization in handling smart cities is a major cause of concern. The cities use the sensors, communication devices and appliances through a variety of vendors and for a variety of purposes. Therefore, integrating the diversified portfolio of the stakeholders becomes a challenging task, posing a threat to proper implementation of AI in smart city development. The application is always better when homogeneity exists. Technology improvements have undoubtedly made life more easy for everyone, but on the flip side, malicious threats have raised serious concerns about how to protect smart cities from potential attacks. In case of Cape Town for example, it was mentioned that "smart city" technologies might significantly contribute to the development of safer, more sustainable and equitable cities, producing models of democracy. Even when such technologies are used properly for the protection and security of the public, there are worries that governance driven by "Big Data" processes heralds the start of a trend of intrusion on the individual's liberty and privacy (Ni Loideain, 2017). Any dysfunctionality of a city management system poses a threat to a range of services offered to smart city residents. For example, if an automated water supply system gets disrupted because of some reason, the supply of water will suffer and so will the residents, as is witnessed at many times. A large amount of information is gathered in the urban areas through smart city networks, and there is always a vulnerability regarding sensitive data of citizens. Data leakage and cross scripting is often experienced in AI-based smart cities (Vogt & Portmann, 2017). Day by day, this issue has been becoming challenging to manage. Due to the ever-growing number of data sets, creating and managing Big Data is a laborious operation that requires data storage, transport, recall and analysis. Due to its significance, big cities face a problem in opening up new channels for managing Big Data management systems.. The revolutionary potential of artificial intelligence to speed up research and development, reduce costs and enhance efficiency has recently been tempered by concerns that these intricate, opaque systems can actually damage society more than they benefit it (Greene, 2018). These projects provide significant incremental gains for their target groups, according to analysis across procedural, rights, instrumental and structural aspects. However, those who benefit most are outside parties and wealthy areas, which raises relative inequality (Kitchin, 2014). Experts in international law, international security, ethics and industry professionals who deal with AI and its applications are concerned about these technical advancements and its threat to humankind (Markotkin & Chernenko, 2020). A major key concern is the environment and its protection. The decline in the carbon footprint has become inevitable for smart cities; therefore, managing things accordingly is a genuine concern in smart cities. The usage of renewable energy is important for overcoming the concern of carbon footprints, so the smart cities are expected to do it for sustainable development. Energy utilization and reduction in the carbon footprint has become the utmost priority of the city development authorities. Another challenge is managing the waste of the city, and it becomes more challenging when the physical structure of the countries is not uniform. On one side, the countries have plains; on the other, the coastal area exists. In addition to this, we also have hot and cold deserts, as well as regions that are more and less likely to see rain, like in India. Lack of uniformity in the physical structure of the countries makes it a very challenging affair in the case of waste management. Also the collection of waste and the segregation of the same are a challenge. Huge populations, unexpected natural calamities and scarcity of resources specifically in the emerging economies are a big hurdle in attaining the main purpose of smart cities. These disasters have a negative impact on the system functioning, and we see breakdown in the systems and failed management. Therefore, we see that the application of AI in smart city development comes with certain challenges and opportunities. The need is to identify and implement appropriate strategies to manage these challenges and get optimal outcomes as far as the positive side of AI in smart cities is concerned. The countries need to work upon recovery strategies to manage the failures in the system and attain maximum output with minimum cost, which is optimization.

In lieu of the same, we will focus upon understanding the opportunities and challenges faced by the Brazil, Russia, India, China and South Africa (BRICS) economies in implementing AI for the development of smart cities.

Kazan (Russia) organized and hosted a two-day discussion forum on October 19–20, 2020, the "BRICS Friendship Cities and Local Government Cooperation Forum."

The discussion focused upon the strategies in order to strengthen assistance and cooperation between the development authorities and municipalities, which would induce the improvement in quality of life and prosperity of city residents. Representatives from 15 countries, 56 cities and 10 major municipalities and local governments participated in the forum. A final declaration was adopted at the finale of the forum where emphasis was on a "Sustainable Development Goals 2030" implementation, which would inevitably bring improvement in the quality of life of people and welfare of the stakeholders (Economic Times, October 20, 2020).

In the trend of urbanization where the rural population is moving to urban areas, as could be seen specifically in the case of Asian and Latin American economies, it is inevitable to understand the dynamics of smart cities.

A long and complicated process is involved in constructing a smart city. The emerging economies have to see environmental sustainability and conducive technologies while implementing AI in development. If we consider the example of India, we see that many cities are starting from scratch and the ones already using AI have many more things to implement. The same is the case with other BRICS economies. The Prime Minister of India stated that in earlier times the cities used to be situated near by the banks of rivers, but presently the setting up of smart cities will look forward to places where the access to fibre optic networks, roads, highways and all-new generation infrastructures will be facilitated.

Beijing, Shanghai, Shenzhen, Hangzhou, Chongqing, Chengdu, Wuhan and Guiyang in China; Surat, Indore, Ahmedabad, Pune, Vijayawada, Rajkot, Visakhapatnam, Pimpri-Chinchwad and Vadodara in India; Obninsk, Moscow, Sholokhov, Yekaterinburg, Kaluga and Samara in Russia; Johannesburg, Durban, Cape Town, Port Elizabeth, Panseria and Mooikloof in South Africa; and Rio De Janeiro, São-Paulo, Belo Horizonte, Vitoria and Curitiba in Brazil have adopted AI at various levels, thriving to be the best smart cities. We observe different levels of implementation of AI in the cities of emerging economies, which was supported in a recent report published by UNCTAD – "Technology and Innovation Report, 2021," which considered 158 countries.

A "country readiness index" is considered in the report, as we see in Table 10.1, which investigates the readiness and progress of various economies in using frontier technologies. Frontier technology includes nanotechnology, IoT, gene editing, blockchain, solar photovoltaic, AI, 5G, robotics, and Big Data. The nations are assessed based on their capabilities in terms of human resources, investment capacity, and technological endeavours like deployment of ICT, financial status, industrial activities and research and development.

TABLE 10.1
Readiness for Frontier Technology Index, Index Score Ranking

Country Name	Total Score	Total Ranking	Score Group	ICT Ranking	Skills Ranking	R&D Ranking	Industry Ranking	Finance Ranking
USA	1.00	1	High	14	17	2	20	2
Switzerland	0.97	2	High	7	13	13	3	3
UK	0.96	3	High	17	12	6	11	14
Sweden	0.96	4	High	1	7	16	15	16
Singapore	0.95	5	High	4	9	18	4	18
China	0.76	25	High	99	96	1	7	6
Russian Federation	0.75	27	High	49	28	11	66	45
Brazil	0.65	41	Upper Middle	73	53	17	42	60
India	0.62	43	Upper Middle	93	108	4	28	76
South America	0.55	54	Upper Middle	69	84	39	71	13

TABLE 10.2
Over-Performing Countries Relative to Their Per Capita GDP

Sr. No.	Countries	Over-Performance Positions
1	India	65
2	Philippines	57
3	Ukraine	47
4	Vietnam	45
5	China	40
6	Jordan	34
7	Brazil	33
8	Republic of Moldova	33
9	South Africa	29
10	Tunisia	29
11	Morocco	29
12	Kenya	28
13	Nepal	28
14	Serbia	25
15	Republic of Korea	24
16	Russian Federation	24
17	Lebanon	24
18	Togo	23
19	UK	21
20	Ghana	20

Source: UNCTAD Technology and Innovation Report, 2021.

Further in the report. the index throws some light on the performance of some over-performing countries. This section of over-performing countries primarily considers the performance of the developing nations. The over-performance of these countries considers the deviation of the estimated index ranking from the actual index ranking, based on per capita income (PCI) of these economies. So the crux is that as per gross domestic product (GDP) of these countries, they are doing beyond expectation. An investigation of the index in terms of outstanding performance revealed that despite comparatively low per capita GDP, they are doing great. With special reference to BRICS nations, four out of five countries make it to the top ten as depicted in Table 10.2.

The index highlights the countries that are out-shining in terms of frontier technologies with respect to their PCI levels. Despite having low PCI, these emerging nations are doing better than expected. The deviation of the estimated index rankings based on their PCI from the actual index rankings is measuring the overall performance of these economies.

The performance of India is best in this category, as the estimated rank was 108 as per PCI but the actual rank holding stood to be 43 so the difference between them stands to be 65 giving India an edge of over-performance by 65 ranks. With 57, the Philippines was the second to make it in this category. Next stands China with 40 and Brazil with 33 ranking positions. South Africa among the BRICS over-performed by 29 ranking positions and finally the Russian Federation by 24 ranking positions was able to make it into the top 20. Among the top ten, four of the five BRICS nations outperformed, which is a matter of great indulgence in using appropriate techniques to transform their economies.

Analysing the rankings on five parameters of ICT, skill, R&D, industry and finance in terms of the BRICS nations, we observe that China is the best among them in finance, R&D and industry ranking sixth, first and seventh overall, respectively. In R&D and industry, the second position is captured by India with 4th and 28th rank overall, respectively. In ICT, Russian Federation and South Africa followed by Brazil stand out to be first, second and third among the BRICS, and in skill rankings Russia, Brazil and South Africa surpass China and India.

The positions of India and China in R&D and Industry are attributed to their skilled workforce available at reasonable cost. The demand generated from a large market presence also contributes to this achievement. FDI inflows are another important aspect and the Chinese government infusing 2 per cent of GDP in R&D does make a huge difference. China and India are the countries which supply abundant labour that is also highly qualified and skilled. This is substantiated by the presence of huge markets promising great demand for products and services, therefore grabbing the attention of multinational giants. In China, 2 per cent of the GDP is invested in R&D and the reflection of this phenomenon could be observed everywhere. However, in ICT connectivity, the BRICS don't show very impressive results and comparing amongst them, it is observed that China and India lag behind Russia, South America and Brazil. In the same manner, they lag behind in ICT skills, especially South America, China and India. The potential of emerging countries of BRICS is immense, and it is of utmost importance to understand the opportunities and challenges faced by these economies on various fronts.

10.2 CHINA

The Chinese government has planned to build 500 smart cities using AI in view of managing pollution, public health, security and traffic. Innovations in AI, ML, 5G, blockchain technology and new energy vehicles have transformed many of the Chinese cities already.

A prominent role is played by the Chinese technology giants who contribute a lot to uplift the modern infrastructure and help in attaining the concerned objectives of the Government of China. Huawei, Didi Chuxing, Alibaba, Tencent and Baidu are some of these examples. In the year 2020, the Chinese government has planned fiscal support worth UD$1.4 trillion for smart city development.

The contribution of these tech giants can be observed in the infrastructural development of various cities. Beijing, Shanghai, Shenzhen, Hangzhou, Chongqing, Chendung, Wuhan and Guidyang are examples of this.as mentioned in Table 10.3.

TABLE 10.3
Application of AI in Smart Cities of China

Artificial Intelligence Application	Tech Entities Involved in AI for Smart Cities
Beijing	
Beijing Citizen Social Service Card – Information related to social security, academics, identity and health is integrated through this service card Robots working in the **AI hot pot restaurant – Robots** carry forward the unmanned system of food ordering, cleaning etc. **Real-time simulation road network model** – The urban traffic is guided here with the help of data on road congestions **Smart museum project** – the main museum's exhibition halls can be projected in 360 degrees with the help of a remote	**TENCENT We Chat Pay** – the digital payment system. In COVID-19 epidemic, We Chat also allowed for continuous updates regarding quarantine measures and provided important statistics to local government officials
Shanghai	
Citizen cloud – this provides services in six categories: health, community life, academics, social security, leisure and transportation. **Framework for traffic information** – this gives information on parking, traffic and public transportation. The passengers are guided in terms of direction with the help of solar sign boards and lids. Face recognition payments for tickets are also on the way **Smart economy** – the focus of development in Shanghai is increasingly in AI, Big Data and cloud computing and related fields	**Baidu and Didi: Intelligent driving and autonomous driving –** Didi Chuxing has developed the Didi **Transportation Brain** and an autonomous vehicle fleet to improve travel times for commuters. Its robotaxi fleet works in coordination with the transportation brain. Yangquan and Hefei have also launched the autonomous driving concept as a pilot programme
Shenzhen	
AI in various fields – the presence of a large number of IT professionals sees the usage of AI in all aspects of the city, which includes education, healthcare, housing, social security and transportation	**Huawei – 5G and IoT** – Shenzhen's Bao'an International Airport signed with Huawei to upgrade airport infrastructure. Due to upgradation, 15 per cent reduction in waiting time and 20 per cent decrease in boarding flight time was observed. Huawei's Smart Traffic Lights and Traffic Control systems help in optimizing traffic patterns in Shenzhen. **Tencent: Net City – Tencent and prominent architectural firm NBBJ** jointly announced the creation of Net City: A concept related to sustainable living, infrastructure focused on pedestrians and a self-driving vehicle network which is underground

(Continued)

TABLE 10.3 *(Continued)*
Application of AI in Smart Cities of China

Artificial Intelligence Application	Tech Entities Involved in AI for Smart Cities
Hangzhou	
Cashless society – Home of Alibaba – unmanned supermarkets, unmanned restaurants and all aspects of life cover mobile payments here **Real-time traffic conditions** are monitored. According to the volume of traffic, the traffic lights change. **Alibaba's AI (artificial intelligence) City Brain system –** is for social governance, security and urban management	**Alibaba**'s City Brain helped in reducing the transit time by 15 per cent. The travel time for ambulances was reduced by 50 per cent and well as traffic violations have lowered. **Alibaba's Medical Brain** employs AI and Big Data for the treatment of patients. Not only have the cities in China have adopted this technology, but it is also imported to Malaysia
Chongqing	
Hub for robot enterprises – various robot research studies, testing and manufacturing take place here, which helps in city management **Urban Big Data joint innovation centre and First National 5G test site** are the USP of this city	**Singapore Telecommunications Limited and China Mobile, China Unicom and China Telecom** are working together on data connectivity between Chongqing and Singapore **Huawei** works on cloud computing, industrial software and sensors; **Rota cloud, COSMO Plat and humi.com** work on information security
Chengdu	
Gigabit network – traffic, education, health, employment and commercial services are managed through an application **Paying with face recognition** is a common phenomenon here	**Huawei** agreed to build an AI-based Big Data centre. Also three platforms for city management are to support Chengdu's urban management, AI innovation and digital economy Urban management platforms will support smart development in fields such as traffic, security, healthcare, contingency management, education and environmental protection. The innovation-supporting platform will help local research institutes carry out advanced technology development Chinese internet search giant **Baidu** joined hands with the Chengdu Hi-tech Industrial Development Zone to promote its self-driving taxi service, the Apollo Go Robotaxi, in Chengdu
Wuhan	
Wuhan has **16 international smart city evaluation criteria, out of which four are used by Singapore and Dubai** for evaluating their projects **A total of 71 application systems** provide services in the field of education, medical care, community services, pensions, transport, traffic management etc.	**Jimu** – this advances a Driver Assistance System featuring collision warnings, monitoring distances, warning related to lane deviation and warning related to pedestrian collision. **Wuhan Zhiling Cloud Technology** – used for high-performance stream data processing, Big Data applications and general enterprise applications **Wuhan Landing Medical Hi-Tech** – pathology analysis system. Automatically identifies cancer cells, precancerous lesions and normal cells. It is also approved for marketing in the US, EU and China

(Continued)

TABLE 10.3 *(Continued)*
Application of AI in Smart Cities of China

Artificial Intelligence Application	Tech Entities Involved in AI for Smart Cities
Guiyang	
Guiyang is considered to be a Big **Data Valley** where companies like Alibaba, Tencent and Apple are establishing their Big Data centres **Tourism brain** – this helps in improving the experience of tourists and better management of the tourism industry	**Guiyang Big Data Group and Yidu Cloud** – Medical Big Data Governance System Co-created by **NASSCOM** and the Guiyang Municipal Government, **SIDCOP** is an AI-powered platform which enables matchmaking between the Indian and Chinese ecosystems to leverage each other's strengths in technology

10.3 CHALLENGES FACED BY CHINA

Alphabet is the parent company of Google. Alphabet's Sidewalk Labs plan to manage traffic in Toronto faced severe protests regarding data privacy and the dominating role of tech giants; in a case that is related to the technology giants in AI and urban development, the scenario is different, as the Chinese government has a prominent role in determining the planning for cities along with financing the projects for the same. In this kind of environment, the people of China can barely give their opinion.

The companies based on AI would find it difficult to export their smart technologies, because according to the US-China Economic and Security Review commission, the US infrastructure could be vulnerable and harm the US interests.

Though the implementation of AI in the development of smart cities has done wonders in the Chinese economy, as every coin has two sides, there is another side of the story too.

The most negative aspect of AI in smart cities is that it brings disruptions in the labour market. The PwC report says that the number of jobs will definitely increase in China, but this will be spread unevenly. Most severely affected labour will be the ones who work in medium-skill and low-skill jobs. On the other hand, those who are highly skilled will get more opportunities (Barton et al., 2017). China is offering specialization in AI to fill the gap between demand and supply of highly skilled labour in AI, but this will have a deteriorating impact as the structural problem remains the same.

A study by the Centre for Strategic and International Studies reveals that by the year 2030, one fifth of jobs in the manufacturing sector will be replaced by automation, posing a challenge for employment. One fifth means almost 100 million workers will be replaced, which is quite challenging. One can visualize these changes, which have already taken place as robots are working in place of humans in several manufacturing units of Guangdong and Zhejiang. The estimate is approximately 40 per cent replacement. Replacement of two lakh workers has already taken place in Dongguan due to deployment of robots. China has a low ranking in vocational training as suggested by the aforementioned report, and if the replacement of low and middle income workforce will keep moving, the inequality in the Chinese workforce will grow inevitably.

The Chinese government is aware of the possible advantages and possible risks and is trying to develop AI strategy on the basis of interaction of these factors (Roberts et al., 2021).

10.4 SOUTH AFRICA

President Cyril Ramaphosa announced that across the country, several new post-apartheid cities are being conceptualized in his Feb 2021 address to the nation.

As innovation is an integral part of AI-based smart cities, the South African economy has initiated AI-based operations in many cities. For example, in agriculture, start-ups such as Aerobotics, MySmartFarm, DroneClouds and in financial services, the Strider-Fintech entity provide platforms that banks and financial institutions use to provide services to new and existing clients. Data Prophet is a South African start-up focused on ML solutions for business. In healthcare, Number Boost manages many mobile HIV clinics. It implemented an algorithm to optimize the clinic schedules and built unstructured supplementary service data for clinic bookings using text messages. Number Boost uses chatbots where patients can clear their doubts, if any. In academia, the University of Pretoria (UP) in South Africa created the Intelligent Systems Group. They aim at creating real-world intelligent systems which are applicable in the South African context. Their main focus areas are remote sensing, digital image processing, telecommunication systems and computer vision.

The University of Cape Town has a Robotics and Agents Lab which conducts research for developing autonomous robots which will certainly play an important role in the coming years with respect to smart city development and especially the overall development of the entire economy. A large number of in-house and foreign entities are working in the area of smart city development and looking forward towards the implementation of AI as mentioned in Table 10.4.

10.5 CHALLENGES FACED BY SOUTH AFRICA

Though there are undoubtedly several positive implications and applications of AI in the smart cities, we find that the challenges are inevitable as the people who are marginalized in the urban areas suffer from different kinds of problems. This has notable implications for social and spatial metamorphosis. There are some major challenges evident in the South African economy. Intelligent modelling tools and Big Data are used to take land- and property-related decisions regarding value and allocation of the same. The satellite images give real-time information and help in analysing the changing scenario of land use in the city. It further helps in urban planning and zoning. However, it was observed that the interests of elites were in top priority and the marginal population suffered on this front. The corporate and real estate entities were actually using those data, and the tendency of monopoly was observed. This was not unanticipated as these segments have access to data and their analytical capabilities help them to use the detailed information in their interest.

TABLE 10.4
Application of AI in Smart Cities of South Africa

Artificial Intelligence Application	Tech Entities Involved in AI for Smart Cities
Cape Town	
Digital government, economy, inclusion and infrastructure are the focus provision of free Wi-Fi zones, and a pioneering Open Data Policy is in practice **Isabelo Smart Bench** – this solar-powered smart bench allows passers-by to access the internet, provides USB charging points and offers a place to sit **CapaCiti**, a job-readiness programme, focuses on skilling youth and makes them industry ready, especially from the areas which lack IT skills such as Gugulethu, Mitchells and Delft **ECAMP aims** to increase transparency and accountability to citizens and entities	**SAP** – a German software company – has implemented their Enterprise and Resource Planning system and some other entities working are **IBM, CISCO, SIEMENS** **Cortex Logic** is shaping AI-based smart development in Cape Town These are AI-driven companies which leverage our unique mix of access to finances, IP, corporates, households, governments, NGOs and the whole community at large
Johannesburg	
900 km fibre optic broadband network – is set up for supporting the communication and broadcasting mechanism. For better management of electricity, **smart metres are deployed. Whenever** the grid will be under pressure, the alert will reach the people and they will manage by turning off high consumption appliances	**Xineoh** – predicts consumer behaviour using huge data bases as well as helps the entities in their marketing campaigns **GotBot** – commercial Chabot platform. The company offers its customer service bots through Facebook Messenger, Twitter, We Chat, Skype, SMS and Web chat
Durban	
ANPR – automatic number plate recognition-camera systems	**MICROSOFT –** AI-based Big Data centres and platforms for city management are to support Durban's urban management
Tshwane	
e-Tshwane – all the stakeholders such as tenants, companies, consumers, households, government interact on this e-platform for all their nee related to everyday urban needs. AI also plays a very important role as they are working on data-driven measures to analyse the consumption pattern in electricity and bring the cost-effectiveness by implementing adequate measures.	**Interfile –** urban management platform will support smart development in fields such as traffic, security, healthcare, contingency management, education and environmental protection
Ekurhuleni	
Video conferencing solution by Huawei – it supports 27 departments and public utilities **Disaster recovery (DR)** – ensures zero data loss and in the case of service downtime, the upgraded data centres ensure a stable operation of Ekurhuleni's diverse smart government applications. One of t hose applications is the **My Ekurhuleni,** which provides access to local government services	**Huawei's** – convergence of communication solutions and cloud data support provide intelligent and flexible connections and help in intelligent digital transformation.

This effectively worsened the digital divide between the elite and marginalized. Contradiction related to use of Big Data was also observed when 15,000 CCTV cameras were installed in Johannesburg by a private entity. Extending surveillance in the monitoring of poorer neighbourhoods is an example of how Big Data and AI-based tools potentially fortify historical sequestration. A "Not Belonging" tag was attached to Black people as they were earmarked in the zones that were surveyed, which is a matter of concern. It seemed that these technologies are resurfacing the apartheid era. In the case of data pertaining to mobility and transport research, it is revealed that a small number of areas are not planned in accordance with the use of applications like Google Maps and Waze, and this can result in new obstructions and spread traffic obstruction into areas that are unprepared to handle high traffic volumes. In the case of data related to mobility and transport research, it reveals that few areas are not planned according to the use of applications like Google Maps and Waze and can engender new occlusion and disseminate traffic obstruction into such areas which are certainly not ready to face large traffic volumes. Therefore, a more integrated approach towards urban transport management is required. Another problem which the development of smart cities in South Africa faces is on the education front. A report on national education policy in South Africa reveals that approximately 18 per cent of secondary school students qualify for higher education; the country is facing an acute shortage of skilled young people, mainly in STEM, which is a critical condition in terms of the implementation of AI and Big Data.

The problem of inclusivity is prominent in the cities of Cape Town and Johannesburg as a wide divide in terms of income and standard of living is evident. A uniform solution will not work here until the government does a lot in the area of providing basic infrastructure and training so that the people could efficiently utilize the services provided by smart technologies. As per a statement given by Cape Town entrepreneur Bradley Shaw, both Cape Town and Johannesburg have to do a lot in terms of smart city development as they are still far from what they should achieve. Moreover, due to increasing migration in South Africa, it will be a challenge to accommodate a large number of people.

For mobile connectivity and broadband facility, the cost for availing internet services is higher than other economies and therefore access to internet data is below 50 per cent of the South African population. In addition, if we look at the global standards, we see that they lack in speed as the connection speed is approximately 3.5 Mbps as per a study conducted by StanLib.

Therefore, we can see that though there is a lot of potential for AI implementation in developing smart cities in the South African economy, the challenges discussed before are major constraints and should be addressed on an utmost urgent basis. Results will be encouraging as the public would benefit at large and the economy of South Africa will tend to be a self-driven economy.

10.6 BRAZIL

The inauguration of the Center for Artificial Intelligence, C4AI, took place recently in the country of Brazil, which will be primarily contributing to the field of AI and its impact on society and economy.

Based in São Paulo, the initial focus will be on environment, education, health, production networks etc., looking forward to inclusiveness, improvement in welfare of people and catering to diverse segments of society. Brazil has always been a pioneer when it comes to smart cities because it began developing them before many other economies in the globe. By the early 1970s, Brazil had reached significant levels of urbanization, with 56.6% of its people living in cities. The period between 1950 and 2000 saw the most intense urbanization. The country's urban areas are home to more than 858 per cent of the overall population. The federal government, along with local urban development organizations, placed great emphasis from the outset on their constructive partnership. In terms of smart cities, Brazil has always been a first mover as it worked upon the same before many economies of the world; high levels of urbanization were achieved by Brazil by the early 1970s as 56 per cent of its population were living in urban areas. The process of intense urbanization majorly took place between 1950 and 2000. More than 85 per cent of the total population resides in the urban locality in the country. The local bodies of urban development in coordination with the federal government emphasize on positive cooperation amongst them since the very beginning. Brazil considers that until and unless there is no remarkable improvement in the lives of city people, the implementation of AI doesn't make any difference. So technological changes and innovative ways are outstanding when in reality the people enjoy the benefits of smart cities. Connected Smart Cities reported government, entities and local bodies work together based on 11 parameters, education, urbanization, access to infrastructure, entrepreneurship, innovation and technology, mobility, environment, health, governance and security to see the progress made by smart cities using AI. Analysis included more than 100 cities and five Brazilian cities were chosen as the best in terms of implementing AI in development. A major cause of concern for Brazil is the concentration of smart cities in specific regions. Deep imbalances could be observed in the economy, specifically the southeast and coastal regions.

The following discussion in Table 10.5 considers the smart cities development scenario in Brazil to know the status of the same.

10.7 CHALLENGES FACED BY BRAZIL

In the year 2017–18 the Global Competitiveness Report by the World Economic Forum stated that in terms of market size Brazil stands out best as compared to other Latin American countries, but when we see Brazil in terms of innovation and technological readiness we observe that other South American economies like Costa Rica and Uruguay are far better than Brazil. In general tax rates, corruption, restrictive labour regulations and inefficient bureaucracy were the main causes of concern and in particular ICT, connectivity and rendering IT services were considered to be the main barriers. The lack of a highly skilled workforce, which fosters creativity and ensures that technology is implemented successfully and to its full potential, is another significant issue in the adoption of AI in smart cities. Approximately 22 per cent of the Brazilian population falls in the category of highly skilled labour according to World Economic Forum's 2017 report, which is far below than Argentina and Chile. For a conducive digital ecosystem, skilled workforce is undoubtedly required.

TABLE 10.5
Application of AI in Smart Cities of Brazil

Artificial Intelligence Application	Tech Entities Involved in AI for Smart Cities
Belo Horizonte	
One of the top 5 smart cities – the city hosts around 300 start-ups and provides a conducive environment which strives for innovation About a hundred Wi-Fi points and a stretch of 700 kilometres of optic fibre are present to drive growth based on AI	**NUCONT** – this manages financial analysis, reporting, accounting, auditing, data analysis, and client management **Logmax** is an AI-enabled TMS (Transport Management System) for the logistics industry. Helps in freight management, document management, real time supply chain management, inventory management, and automated alerts. **Dunning** offers a debt collection strategy suggestion platform to lenders. They can identify and prognosticate payment capability of borrowers. Use of machine learning, data mining, and Big Data optimization algorithms helps in generating credit indicators and collection strategies
Vitoria	
Trabalha Vix (Work Vix) provides formal employment offers to the residents. It has trained many entrepreneurs in various fields. **Ponto Vitoria (Vitoria Stop) and Minha Consulta (My Appointment)** are some other examples of usage of AI in smart city development in Vitoria	**Vale** aims to increase productivity and operational efficiency, achieve the best health and safety rates, improve financial performance and drive innovation for betterment of people in health, transport, education, security, environment, logistics, e-commerce and much more
São Paulo	
Descomplica, which means uncomplicated, provides more than 300 government services to the residents which enhance the quality of life. Services include unemployment insurance, parking cards for the disabled and elderly, military service registration, issuance of municipal taxes, and others. Additionally, the initiative is further extended into Descomplica Digital, which furthermore contributes through enhancement of services to residents	**Companies using artificial intelligence (AI) such as MEMORELAB, B R Q, U O L DIVEO, ENDEKEN and IQVIA are aiming to deliver information about social security, education, transportation and health to urban residents**
Rio de Janeiro	
The COR – City operation center aids in enhancing utility and agency mergers. The goal of the Digital Inclusion initiative is to engage the locals. In transport and mobility, exclusive transport corridors are introduced with the **Unified Ticket** system The air quality of the city is monitored by MonitorArRio Program and is highly valuable as it meticulously identifies emergency situations and warns the residents in advance. **Digital Traffic** monitors and reports in real time.	**IBM's Intelligent Operations Center**. Contributes with real-time city information by using a forecasting system that synthesizes data from the river basin, topography surveys, historical rainfall logs, and radar feeds. The operations centre is able to anticipate heavy rains, flash floods, landslides, power outages, and traffic hazards **CARIOCA DIGITAL AND 1746 RIO PREFEITURA** are a few more names which are helping the city in terms of implementing AI for better lives

(Continued)

TABLE 10.5 *(Continued)*
Application of AI in Smart Cities of Brazil

Artificial Intelligence Application	Tech Entities Involved in AI for Smart Cities
Curitiba	
The creation of smart cities utilizing AI is being promoted by innovation hubs like Vale do Pinho um Bom negócio, Fab Labs, urban farms, and apps that assist inhabitants in a variety of areas. Curitiba is promoting a "smart city movement" to build an innovation ecosystem to promote smart solutions that are aligned with the implementation of the SDGs The city uses 70 indicators divided into 11 axes like environment, education, mobility, energy, economy, innovation, education, health, safety, technology and governance. It ensures that everyone's welfare as well as impact assessment can be done	**Offerings from Phone Track comprise tracking, monitoring, campaign measurement, call scoring, digital IVR, platform integrations, and voice analytics. Smarthint offers product suggestions based on the user's profile, pop-up retention, customer behaviour analytics, and other factors. Through its dashboard, it gives consumers analysis reports. It also offers advice for marketing strategies via webinars and e-books. Through warehouse management interface with ERP management systems, warehouse automation, product traceability and digital farm operations, the Fredbots chatbot platform, agro top, aids in agro-management. Phone Track**-offerings include tracking and monitoring, campaign measurement, call score, digital IVR, integrations with other platforms and speech analytics. **Smarthint**-provides product recommendations based on the customer's profile, retention of pop-ups, customer behaviour analytics, and more. It provides the analysis reports to the users through its dashboard as well as provides guidance for marketing campaigns through webinars, e-books. **Fredbots** – Chatbot platform, **Agro top** – it helps in agro-management through warehouse management integration with ERP management systems, warehouse automation, product traceability, and digital agriculture operations

Increasingly, companies seek a skilled workforce to integrate and develop a digital ecosystem. According to the United Nations Conference on Trade and Development, the lack of high-tech companies in Brazil is also a major cause of concern in successful implementation of AI in smart city development.

In cultural and societal aspects the challenges of maintaining human interconnection and a sense of belonging are of utmost importance. The GPS based tracking data, associated with data on consumers regarding preferences, location, and personal interests based on the IoT, poses security and privacy concerns. Reinforcement of existing social inequalities could be observed in the country rather than observations on the positive side where inclusiveness is witnessed. There are populations which do not wish to interact with the digital infrastructure of smart cities because of privacy concerns. Also risk in terms of unequal growth is inevitable when the affluent society segment has a dominant role to play in comparison to the peripheral segment. Massive development that has been brought up by smart cities has also posed a threat to the environment. Migration to urban areas from rural areas is causing ecological stress and pollution problems. Unplanned settlements in the urban areas are responsible for congestion and groundwater levels are suffering due to this massive development.

Lack of citizen participation is also considered to be a major challenge in smart city development in Brazil. Most of the on-going projects have no participation from the public and therefore the representation is almost negligible from the locals who actually face challenges on a day-to-day basis. Monolithic policy formulation induces ambiguity and better views could not be placed for smart city development.

10.8 INDIA

Recently centre released the list of smart city winners in different categories for the year 2020.

On social aspects, Tirupati bagged a Health Benchmark for Municipal Schools; Bhubaneswar in Socially Smart Digital Library Solution stood to be first. On the cultural aspect Indore bagged for conservation of heritage, Gwalior for digital museum. In terms of urban environment, Bhopal received clean energy and Chennai for restoration of water bodies, Tirupati for renewable energy generation. On the sanitation front, Tirupati scored for bioremediation and bio-Mining, Indore for municipal waste management system and Surat for conservation through treated wastewater. Indore was praised for its Carbon Credit Financing Mechanism. In terms of boosting local identity and economy through Design Studio, Agra was praised for its micro skill development centre. On water conservation front Dehradun for its smart water metering and water ATM and Varanasi for its Eco-Restoration of Assi River, Surat for its Integrated and Sustainable Water Supply System, stood out in terms of smart city rankings. In transportation, Aurangabad with its Majhi Smart Buses, Surat with Dynamic Scheduling Buses and Ahmedabad with man-less parking system and automatic ticket dispensing machines bagged their respective rankings. India aims to develop 100 smart cities and the selection process will consider the population residing in urban areas and statutory towns in Indian Union Territories and states. Shillong, the capital of Meghalaya was included as the 100th city in June 2018.

The smart city mission was launched in June 2015 under the leadership of Prime Minister Narendra Modi. A better quality of life and sustainable environment are the main agendas of the government of India in terms of usage of AI in smart city development. The infrastructural elements under smart cities were a well-connected, efficient public transport system, constant quality supply of water and power and ample opportunities for jobs and livelihoods focusing on safety and security, health, education, energy efficiency, electricity efficiency, recycling of waste and water conservation, telemedicine, tele-education, skill development, citizen monitoring, crime management, grievance redressal, security and surveillance and many more. Also the government has invited discussion on smart parks and public facilities, smart homes, crowd management, and intelligent safety systems.

In India, unlike other countries, the smart city mission are area specific missions and are complemented by other project specific missions like, AMRUT, HRIDAY, Housing for All, Skill Development, Digital India, and Swachh Bharat Mission – SBM.

"AI for ALL" is what India is looking forward to. India has the advantage of AI-trained workforce which is visible in various AI start-ups coming up with wonderful solutions for smart city management and many more avenues; therefore, the nation can definitely look forward to AI-driven solutions and bring a revolutionary change in the areas of education, healthcare, manufacturing, transportation, security, services and agriculture. When compared with their counterparts, especially China, we see that India is in the initial stage of implementing AI-driven solutions and the projects are to be completed by greater volume. Based on a survey by the Ministry of Housing and Urban Affairs, Government of India under the Smart City Mission, the following projects of smart city development using AI were observed as mentioned in Table 10.6

Coimbatore, Kakinada, Belgaum, Udaipur, Guwahati, Chennai, Ludhiana and Bhopal are the cities of further ranking, respectively, where the implementation of AI is in progress with several projects going on.

The young, large population of India is always eager to adopt AI and accordingly they are training themselves also. On the Stanford AI Vibrancy Index, India has bagged the second position and the biggest contributing factor here was the AI-skilled and properly trained workforce. Thousand lakhs of students get trained every year from the Indian premier institutes, like Indian Institute of technology, National Institute of Technology, Indian Institute of Information Technology and so on. These leading institutions have generated a plethora of highly skilled workers and have also been a ground for incubation and several successful start-ups.

Deep-learning algorithms based on ML in AI can give discernment to the healthcare industry in predicting any future worries of patients. Early detections and prevention of the same will be a big relief for the same. In Bangalore, one of the start-ups had developed an AI-based technology to screen for initial signs of breast cancer and in Tamil Nadu, ML algorithms are applied to find diabetic retinopathy, as they are trying to face the challenge of shortage of eye doctors. Throughout COVID-19, an AI-based Chatbot was used by MyGov, so that proper communication can be maintained throughout. Similarly, the Indian Council of Medical Research used the Watson Assistant so that proper communication between the stakeholders could takes place. Initially the pharmaceutical companies used to take longer days and months for their clinical trials, but we see that with the application of AI, the timeframe has come down to fewer days.

In the field of agriculture, we see how a big role is played by AI-enabled technology and processes as we see various solutions on pest control, irrigation, soil conservation, crop insurance and water management. These are helping the farmers in all spheres and contributing towards better crop production and management of the same. To give support in weather forecasting so that proper decision can be taken in time, a pilot project based on AI has been launched in the states of Maharashtra (Nanded), Madhya Pradesh (Bhopal) and Gujarat (Rajkot). ICRISAT has created a power-sowing software based on AI that encourages farmers to forecast weather conditions and provides them with timing recommendations for when to plant their

TABLE 10.6
Application of AI in Smart Cities of India

Artificial Intelligence Application	Tech Entities Involved in AI for Smart Cities
Bhubaneshwar	
Smart Water Meter and **SCADA** Smart Waste Management Smart Environment Management, United Nations Sustainable Energy for All **SE4ALL** Project, Traffic Network Flow, Monitoring Extreme Weather Disaster Response are some of the AI applications in the city of Bhubaneshwar	Bhubaneswar is managed by **Infosys, Tata Consultancy Services, Tech Mahindra, Mind tree, and Wipro,** in terms of providing IoT and AI-based services
Pune	
Smart parking, smart metering (water) electricity distribution – smart grid and metering, e-gov IT connectivity, adaptive traffic control system Bus system **ITMS**, Pilot **DMA** for 24×7 water, helium leak identification Smart commercial meters	**Pune-based ALTIZON** – Altizon provides an industrial IoT platform for manufacturing applications. **Shoptimize** provides cloud-based customer engagement solutions for e-commerce. **NLPBots** is a chatbot development service that builds and integrates chatbots for multiple enterprise clients and for multiple use cases. **Zinrelo** offers a loyalty programme platform and a referral platform
Jaipur	
SMART mobility – rickshaw stands, intelligent **IPT** (auto and taxi) stands Multimodal integration at UG Metro stations, **Digital Signage** for traffic and tourism Intelligent Traffic Management System – **Synchronized Traffic Signals** for optimum traffic flow, **Smart toilets** – coin-operated entry, self-cleansing, solar-powered, **GPRS** Connected with display boards and voice guidance. Deployment of **SWM** Operations systems and asset management systems	**Epista**, a Spanish company and **Lee** (South Asia) have proposed to execute the project for developing a smart city in Jaipur. **Tata Consultants, Webcos and Sai Consultants** have proposed to execute smart city projects both in Udaipur and Jaipur
Surat	
Novation of **STPs with SCADA** and energy generation, smart street lighting (LED) and monitoring system, innovation, incubation and start-up and Trade Facilitation Centre, Advanced Grievance Redressal System. **Smart Waste Collection System** Air and water quality monitoring system Area Surveillance Network, **Surat IT-MAC** (Integrated Transport-Mobility Administration Center) **Automatic Fare collection system** SMAC Center (Smart City Center) **MySurat.in** [Active Citizen Engagement] S-Connect Card Management System **Connected Surat** Wi-Fi-Surat: FTH-Fibre to Home	**SMAC** Center is envisaged as an administrative control centre for the city of Surat for effective and efficient delivery of all civic services. Surat Smart City Development Limited (**SSCDL**) is formed as a special purpose vehicle (SPV) for the implementation of the smart city projects at the city level **Microsoft City Next** has partnered with Surat Municipal Corporation (SMC) to develop Surat as a smart city with advanced urban planning and citizen empowerment processes never heard in India **IBM** picks Surat, Allahabad and Vizag among 16 global locations for smart cities project.

(Continued)

TABLE 10.6 *(Continued)*
Application of AI in Smart Cities of India

Artificial Intelligence Application	Tech Entities Involved in AI for Smart Cities
Kochi	
Smart Wi-Fi bus shelters with LED bulbs, FM radio, USB drive slots, magazine/newspaper kiosks, mobile charger units, **Drone Surveillance, Spatial Mapping of assets and SCADA systems for Information gathering with monitoring** centres	**Smart City Kochi Infrastructure Private Limited** in collaboration with **GEMS Education, MARATT, Prestige Group and Infra Band are** providing AI-based services in various spheres
Ahmedabad	
Smart Waste Segregation Smart Metering – OFC Network Wi-Fi Hotspots **NMT Tracks** and Cycle Tracks Intelligent Traffic Management; **Smart Parking** surveillance cameras, integrated transit management platform; **Common Card Payment System** (smart cards), readers, software, **GPS and AVLS, PIS**	Waste management solutions company **Envac** is working on sustainable waste management in GIFT city in Ahmedabad **Smart City Ahmedabad Development Limited (SCADL) is** for the implementation for the projects under the Smart City Mission for the city of Ahmedabad
Jabalpur	
Multi-level parking with smart components, intelligent traffic management system, **SCADA** Sensor-enabled flow metres SCADA-enabled pressure transducers SCADA-enabled level sensor system **Geo-Thermal Imaging Leak detectors** Smart Consumer Water Meters Software for area Command and Control Center Water Quality Monitoring System Sewerage existing Network laying of new sewerage network Sewerage Primary Network for **DEWATS**, Smart Components in Power Substations Smart Components in Substations, **GIS Mapping** of the Area linking all the properties, services and its management and monitoring, **Air Quality Monitoring Sensors** on unified pole	Delhi-based infrastructure consultancy firm **Rudrabhishek Enterprises Limited**, for the implementation for the projects under the Smart City Mission for the city of Jabalpur
Visakhapatnam	
ICT-enabled **central command centre** capable of communicating with other city systems such as early warning systems, emergency response forces, utilities (energy, water and sewage SCADA systems), transportation, city surveillance, citizen engagement platform **Domain solutions related to disaster management** such as risk assessment and modelling system, early warning system, emergency response forces tracking through GPS solutions, communication system	**Salasar –** smart city poles, camouflaging solutions, **Cyient** – Cyient's clients range from defence and healthcare to rail transportation and energy solutions. **Fluentgrid –** a smart energy and smart city technology solutions provider, **Prospecta –** created to power Master Data Governance and business collaboration, Prospecta aims to be top of mind when one thinks of data safety

(Continued)

TABLE 10.6 *(Continued)*
Application of AI in Smart Cities of India

Artificial Intelligence Application	Tech Entities Involved in AI for Smart Cities
Collaborative platform for on-going citizen engagement is strengthening city services systems such as property tax, healthcare, utilities, city surveillance, transportation and bringing them to the central command centre	
Solapur	
Mapping of utilities and implementation of **SCADA** Vehicle tracking and Monitoring System, **smart meters, smart parkings, e-toilets, street sweeping machines, 4630 smart poles**, Junction improvement with **CCTV and ITS systems for 16 junctions**	
Davanagere	
GPS- and RFID-Based Integrated Bin Less Solid Waste Management System, Intelligent Solar Street Lights, **Underground Utility Corridor with sensors**, ICT-enabled E-Governance System on GIS, **Intelligent Urban Mobility System**	
Indore	
Smart Parking and Other Components – integration with Pan City, Geo-Thermal imaging Leak detectors, **Smart Distribution Network** with Meters and sensors, Multi-purpose Communication network: laying of **multi-purpose OFC communication network** with a bandwidth of 10 Gb and 25 km range **City Dash Board**: collection, collation, analysis and dissemination of information for all	**Indore Smart City Development ISCDL** – this urban management platform will support smart development in fields such as traffic, security, healthcare, contingency management, education and environmental protection
New Delhi	
E-governance and Computerization in ULB. In various fields, such as property tax accounting, water supply and other utilities E-governance are implemented. Birth and Death Registration and Health programmes, Citizens Grievance, Monitoring Personnel Management System, Building Plan Approval, e-Procurement are few steps towards AI applications in smart city of New Delhi.	New Delhi–based **Quytech** develops Mobile Apps, AR/VR, AI, Game, **Acropolis InfoTech (P) Limited** – an award-winning App/Software Development Company, **Valiance Solutions –** driving business excellence using AI and ML. **OnGraph** is a global software development company that has rendered quality web and mobile development services to companies engaged in varied markets like healthcare, banking, finance, ecommerce, education, telecom, media and entertainment and more

seeds.. Results are quite good as an increase in the production is found threefold. Financial inclusion is also being worked on by AI-based fintech companies so that the rural population is not deprived of the financial facilities.

An AI-based model has been implemented in the state of Bihar which is used for flood forecasting. It is a big sigh of relief as it ensures the safety of more than 200,000 square kilometres and about lakhs of people residing in that area. The warning signs alert the entire region 24 hours prior, and the best part is that the alerts are communicated in nine local languages, which ensures that the communication is proper and in time, which saves so many lives.

The Government of India and its various ministries are also making plans and implementing them so that AI-based training can be given to various segments of the population. For example, the CBSE has introduced the basics of AI, ML and data science from the sixth standard itself, which ensures that students passing out have the basic knowledge and skills of data science, ML and AI. "Responsible Artificial Intelligence for Youth" is a course launched by the Ministry of Electronics and Information Technology, and thousands of government school students received their training in basic AI and ML.

To prevent use of unfair means in the examinations AI has come to the rescue of the education sector and is also used to track the students' performance and prevent student dropout.

In the area of transportation, AI-enabled systems in public transport, in the states of Uttar Pradesh, Karnataka, have been very useful to avoid collisions in order to alert sleepy drivers. Two sensors are in the AI backed anti-collision systems.

Despite implementing AI-based smart city development, there are several key challenges that are faced by the cities on a micro-level and the entire nation on a macro-level.

10.9 CHALLENGES FACED BY INDIA

The pace at which smart city development in India is taking place is a major cause of concern. Only 50–60 per cent of projects have been completed since the initiation of the smart city project, which is an indication of slow pace of development. When we analyse various aspects, we find the shortcomings on both the management front and technological front, which also indicates the societal challenges that the citizens are facing.

In management, we see that the Special Purpose Vehicles are not functioning properly. For example, in the state of Goa, in Panaji, the Special Purpose Vehicles were functioning in the absence of boards over a year. Several violations related to audits were observed. Difficulties in understanding huge data sets and extracting proper information from the data sets have also been a major area of concern. Though the Ministry of Housing and Urban Affairs, in a joint effort with Tata trusts, has conducted development and training programmes to train the stakeholders, slow results are witnessed. The same was seen in the case of other cities as well; in the case of Ludhiana, the director of LSCL-Ludhiana Smart City Company Ltd expressed his dissatisfaction over execution of projects. Lack of coordination among various government departments is also a serious concern for smart city projects. Another aspect of the challenge is the financial aspect. Efficient use of funds and transfer of funds from the state and local governments to special purpose vehicles is a big hurdle. For example, in the case of Srinagar, the amount allocated from the centre was approximately INR 1.27 billion, but the release of funds from the state and local government stood to be only 200 million. Sometimes financial crunch is also a matter of concern; in the case of Faridabad smart city development, it was observed that the Faridabad Municipal Corporation's financial condition was not viable. Out of more than 25 projects, only few have been completed. Delays in release of funds are also an area to consider. No matter how much money is sent from the centre to the states, it frequently happens that only a portion of the money is released, which results in only a partial development of the cities.

Smart cities are also facing the challenge of data security. Network-connected devices generate ample data, but hacking of confidential data and illegal usage of the same are again a threat to the communities and the stakeholders. Adequate data protection techniques are an area in which more and effective efforts are needed. The hacking in February 2021 in Oldsmar, Florida, USA, is an example of how hacking can have major serious and devastating effects on the nation. Here the hacker increased the sodium hydroxide level in the water supply, which resulted in serious health concerns. This has never happened in India but if effective methods to prevent hackers are not brought on board than, this can happen anywhere, damaging the lives of citizens. Instituting the notion of smart cities in India is a good idea, but because India sufferers in basic infrastructure and basic amenities, many challenges have to be faced by Indian cities.

In India, mass migrations to the urban areas are responsible for the crunch in the basic amenities, leading to various problems, like overcrowding, sanitation, open spaces, electricity consumption, densely populated areas, pollution and other kinds of burdens on the available resources. Over-utilization of resources always leads to issues of effective productivity and hampers the opportunities and pace at which these cities may develop and grow.

These issues need to be addressed in topmost priority so that the dream of converting cities into 100 smart cities becomes real. Apart from several initiatives, the Government of India is also working on the private-public partnership model to develop smart cities and overcome these challenges. The Indian government is concentrating on the information technology and ITES industries and is attempting to attract foreign direct investments from significant global firms. Major governments have expressed interest in India's Smart City goal, in addition to notable actors throughout the world who have a clear interest in India.. Major funding comes from the United States of America, Spain, France, Germany, Japan, Sweden, Singapore and many other countries. The goal of the Indo-Canadian collaboration, which has 20 cities in the states of Haryana, Rajasthan, and Punjab, is to complete smart city projects as quickly as possible. The development of smart cities in India is the result of numerous collaborations between the Government of India and various AI, ML rich technology countries. Their development agenda includes proper training for the officials and localization of services for inclusivity purposes. So the real challenge that lies ahead is to include each and every segment of the society and the economy and to make them sustainable where the residents are at ease of living and people are trained to use the resources wisely and to maximize productivity.

10.10 RUSSIA

In the current scenario, the quality of life, social development and economic growth along with international competitiveness are determined by the level of technological development of any country. The Russian economy is also striving for the same. The

President of Russia on many occasions deliberated on the importance of AI and smart cities development and spoke on many occasions on the relevance of studying and developing AI. In lieu of the same, the government of Russia is working on various strategies to use AI widely in all spheres of life. A national programme on "digital economy" was initiated and the big cities in the Russian economy are working on various projects where the usage of AI can be incorporated in the area of waste disposal and management, traffic and parking management, security and safety, healthcare, transportation, education, monitoring and many other aspects of the smart city. Kazan, Moscow and Saint Petersburg are considered to be the smartest cities of Russia. There are ample opportunities for smaller cities of Russia to evolve into smart cities, and the government is also working upon it but it will take some time.

Russia is a particularly interesting case study for smart cities, because a large portion of the population is concentrated on large urban areas, and a smaller part resides in rural areas. In this case if smart cities are developed which are AI-enabled, we can assume that the majority of the population would get advantages from the advancement in the processes, as we know that in European Russia 70 per cent of population resides in 20 per cent of the area, and we observe that majorly the population reside in Moscow where approximately 14 per cent of the population reside; secondly, the population reside in St. Petersburg with 7 per cent of population, followed by Kazan and Yekaterinburg.

Saint Petersburg, Moscow and Kazan are the cities in Russia which have mostly used AI in development. In the area of transport, it was found that the traffic lights smart management in Voronezh decreased the time of crossing the road by 18 per cent. In Russia, the fast-paced development of public transport systems is taking place. Major cities have already implemented smart tracking systems and smart payment systems. Four hundred and seventy-five smart buses are being operated in Moscow to safeguard the environment. Green transport is what they are looking forward to in future.

Smart communication is also a technique in which the cities of Russia are involved. They have a system of unified portals to render public services called Gosuslugi. This urban management platform will support smart development in fields such as traffic, security, healthcare, contingency management, education and environmental protection. SMS notifications related to adverse weather conditions and emergency situations are received by the people of Moscow.

In smart services, various smart applications to education sector, healthcare and surveillance and security are being used, and people have benefited from it. Smart classrooms, video conferencing, online appointments with doctors and surveillance cameras are frequently used and the life of people is becoming very productive and easy day to day. In the education sector, the smart devices have enhanced the communication between the students, teachers and concerned authorities.

When it comes to smart environments, various safety systems are installed, for example automatic fire detection systems, identification of persons entering the premises, intercoms and systems for fingerprints, which have made life secure for the residents.

In the area of agriculture, automated systems are helping in all spheres. Automated disease detection, anti-pesticide sprinkling, soil conservation, opening and closing of

TABLE 10.7
Application of AI in Smart Cities of Russia

Artificial Intelligence Application	Tech Entities Involved in AI for Smart Cities
Moscow	
Our City is a platform to receive grievances, responding to the complaints and giving solutions to the residents on an urgent basis. **Active Citizen** provides a platform to residents of Moscow to raise their voice and give suggestions related to the transport system, development issues, safety concerns and other services. **Crowd** is an attempt to expand democratic processes and encourage citizens' involvement enabled by Artificial Intelligence. It involves the citizens and welcomes their ideas on the city's development. **My Street** is related to development of parks, community halls, streets and open green spaces to give a pleasant experience and joyful accessibility to the citizens. It also looks after the renovation of parks, streets and any required spaces. UMIAS- **Unified Medical Information Analytical System** is used for medical care for urban population	**Beeline** is a Russian telecom entity which provides service on a narrow band IoT network and provides connections to more than 10,000 devices. **Botkin.Ai** is an entity working in the field of medical devices using AI. It has enabled medical experts to detect the lung cancer up to 95 per cent accuracy, and in the case of pneumonia, the detection is approximately 99 per cent accurate. **Profilum** is a platform which is providing useful insights to school students by helping them through counselling on their future choices in terms of profession. Moreover, it is updating the authorities about urban infrastructure according to the needs of youths to empower them in building up their careers **Genotek** is related to genetic science and is contributing at a large scale in terms of DNA isolation, bioinformatics, genetic testing, clinical trials and disease diagnostics
St. Petersburg	
Park Mobile is a smart parking facility where residents use technology-based services. **5G internet** facility and **cellular technology**, the highest-powered connections, are run via **fibre optics**. **AI-based services** like healthcare, banking and finance, e-commerce, education, telecom, media and entertainment and more are part of this smart city development	**Duke Energy** and **Spectrum** – an AI Alliance of Russian corporates. Also the internet giants, such as **Yandex, Mail.ru Group and Gazprom Neft Energy, MTS**, are the entities working upon developing AI-based services which will make life easy for the residents of the city

vents, automated recycling, targeted measurements and automated irrigation make the crop production affordable and properly available. Targeted measurements make things simple and at all macro-levels ensure food security.

A number of AI applications are witnessed in the Russian economy as well as depicted in Table 10.7.

10.11 CHALLENGES FACED BY RUSSIA

Until November 2019, Russia lacked national policy and strategy in AI and this is the reason why the Russian Federation failed to emerge in any world ranking in terms of AI competitiveness. After the country came up with the national policy on the same, we see that there are two time frames on which the country is working with respect to AI. The first emphasizes the development and implementation of AI in Russian cities by the year 2014, and the second one emphasizes on improving its position in AI in the world technology and innovation index by 2030. According to the **Readiness for Frontier Technology Index 2020**, the Russian Federation ranks 27th, which is better than their Indian, South African and Brazilian counterparts and a rank gap with respect to the Chinese economy. This indicates that the government of Russia is quite serious about its innovation and technology projects and is very much concerned with equipping the cities of Russia with AI. The national policy clearly indicates the use of AI in export-oriented industries, particularly in agriculture and manufacturing. Apart from this, Russia is also targeting to increase the technology-based entities by 50 per cent in near future. All these things will definitely make life easy for the residents and the use of AI in health, education, banking and finance, transportation, security and safety, transportation, traffic management, automated industries and many other aspects. However, Russia is trying and doing well, but the scope of implementation of AI in the Russian cities is huge. Only a few cities are using AI in an actual sense and much more needs to be done in the majority of cities to make them smart. It is likely that a long-term strategy is required to achieve the goal; therefore, it is important to understand what aspects the Russian Federation are lacking when it comes to smart city status in Russia implementing AI.

The climate conditions in Russia are a big hindrance for an impressive network of public transport. The smart technology is not easy to use in winters when the cities are covered with snow and chilling low temperatures. Also the residents are used to having their own vehicles, and congestion becomes a big problem. According to reports, the penetration of vehicles on the roads of Moscow has increased by many folds; therefore, the smart system does not operate in an efficient way. Problems related to waste management are also very common in Russian cities. Smart garbage segregation solutions and collection of garbage are a big problem; therefore, the aim of a smart environment could not be fulfilled. The problem not only lies in the proper implementation of the concerned smart solution but also in the people who are not properly trained to do the same. Dissemination of information and training lacks in usage of AI-based solutions. Smart environment also has a problem related to fewer numbers of green charging platforms. Proper installation of charging electric vehicles is lacking; therefore, the cities face the challenge of making the environment conducive and green.

Dissemination of information and making the public aware of various aspects of AI used in the cities is an area which needs to be worked upon. For instance, a proper utilization of smart waste management and green technologies can properly be done only when people are aware of its utility. More action and strategy from the government along with advertising campaigns will give the solutions. Population ageing is also a prominent challenge faced by the city of Moscow, as according to the OECD report, approximately five lake residents will reach the age of 80 years in coming 10 years of time. The city of Moscow will have to address and find solutions to population aging in the near future. This challenge in the demographic structure will induce several pressures. Along with these challenges, the balanced urban development is a major cause of concern. Imbalance development in urban aspects is not only a challenge itself but will also bring in ample related challenges. The smart city development also needs to pay attention to SDGs and give solutions to various trade-offs taking place for sustainable development.

10.12 POLICY IMPLICATION AND CONCLUSION FOR BRICS ECONOMIES

In all nations, the technological changes bring in inequalities in social and economic structures. To take advantage of frontier technologies, countries must not only promote the usage but also adaptation and adoption, which will help in bringing down the adverse effects of the same. When we specifically talk about the developing countries a few challenges are inevitable. Technological disparities, demographic changes, financial crunch, issues related to intellectual property rights, job loss due to automation, polarization in job, fragile labour rights, parallel economy, concentration of profit and inequalities in income are the structural problems which these economies face. These challenges increase when AI-driven development induces the divide more.

The national governance and conducive policies, collaborations and cooperation at international levels, and foremost important active participation by citizens of the nations and their appropriate training are the most vital issues to overcome the challenges of AI. Special attention is required in terms of technological changes which are taking place at a very fast pace. The social structure and the need of the society are to be borne in mind to take optimal advantage of technology-driven development. UNCTAD urges developing countries to brace their innovation structure, as they are weak and vulnerable to systemic collapse and structural inadequacy. Dr. Shamika N. Sirimanne, Director, Division on Technology and Logistics, United Nations Conference on Trade and Development (UNCTAD) stated, "A whole-of-government approach is needed to absorb these technologies, as opposed to working in silos."

Alignment of science, technology and innovation policies, along with industrial policies, export-import policies, fiscal-monetary policies, education sector policies and policies related to other sectors is a must. The new technologies have spark to increase the productivity of the economies, which can enhance the speed of already existing traditional production bases and bring a massive structural transformation. The inclusion of STI policies in the national development plans would strike a balanced approach to overcome the problems. Training of workforce and the stakeholders so that they conceive the required skills is also necessary for conducive development. For digital inclusiveness and digital literacy, the infrastructure also needs to be worked upon to provide them the basic amenities, so that the people could be well connected. The government should provide digital connectivity services at a subsidized rate. Various incentives should also be arranged to work on inclusivity.

The major area of concern for BRICS is to implement AI in an efficient manner so that the proper utilization of resources and effective development of cities in these economies takes place. The BRICS nations are working on the gap areas and are trying to best utilize the AI in development of their cities. These countries are working on surveillance and security, aerial monitoring, traffic management, vehicle parking, waste collection and waste disposal management, planning management and city administration, energy consumption optimization and many other areas. They are also strategizing for managing challenges in the area of hardware installation, high maintenance costs, security issues, privacy concerns and discrimination in society on various aspects.

Security concerns and privacy issues are two of the major concerns that the economies are facing at present. Though various policy formulations are taking place to overcome their concerns, when we look into the data, we find an ever-increasing tendency of criminal activities like robbery and theft. The use of AI-backed surveillance systems is contributing to managing these issues with the help of face recognition, suspicious behaviour alert, license plate recognition, traffic misconduct alert and many other things. Not only will AI networks capture criminal activity, but they will also provide authorities with information that will enable them to be aware in time and reduce the likelihood that any crime will occur. Another aspect of AI playing a conducive role is aerial monitoring. Tremendous growth in the area of aerial monitoring is seen in recent times. Surveillance cameras attached with drones are helping to authorities to watch places of concern like gatherings at events, borders, and concerts, and are helpful to authorities in fulfilling their missions. Any kind of suspicious activity is detected, and the stakeholders are alerted with the details regarding location. AI is being used by several governments to map out areas in order to locate uncharted territory, and this has a significant impact on the plans for urban development of the country.

Face detection through AI is preventing fraudsters from committing fraudulent activity. AI is also used to seek for missing people. Any unauthorized access is denied, which is a big sigh of relief in the digitalized world.

The use of AI in traffic management and vehicle parking is also very important. The problem of parking space and traffic jams is a big challenge in traffic management of the cities. Smart traffic signals are examples of AI in traffic management which manages the flow of traffic in a proper way. Reduction in travel time and consumption of fuel are also brought down. Infringements are also captured through the AI-enabled traffic management systems. ML and AI also help in detecting if parking spaces are available in the parking area. Payment for parking of vehicles is also done through AI, which saves time and fuel.

In addition to smart cities, another major area, in which AI is contributing, is waste management. Collection and disposal of waste is a very important area of consideration in smart cities. Managing garbage collection is a big challenge for smart cities. Due to urbanization taking place on a large scale and migration of people from rural to urban areas, the population density is increasing day by day. The pressure on resources of the cities is known to everyone. Densely populated cities generate high quantities of garbage also. Efficient waste management systems can only solve the problem and we see that AI has a big role to play in this regard. The usage of AI has helped waste collection and disposal of waste. AI-enabled sensors installed in the garbage cans notify the authorities about the spaces that are filled with garbage and the spaces that are left to accommodate more of it. This helps the garbage collectors to form a path accordingly and save time and fuel. AI not only helps in the collection of garbage, but the sorting is also done based on the fed software instructions. Therefore, we see that all this has helped one to a large extent, and the management of smart cities becomes easier with these AI-enabled processes.

AI is also vital for the administration of cities and has helped a lot in planning smart cities in BRICS. The data related to demography of the cities makes it easier for the planners to implement the policies of welfare in an effective manner. This information on city demography helps the planners to use the resources properly and the administrators access the data accordingly. AI also helps in planning for expansion as the AI-enabled technologies help the authorities to tap the missing land which can be used later for the eventual expansion of cities. Also the infrastructural development gets lots of help from AI-enabled systems.

The power and water supply is vital for all cities. The usage of AI in BRICS cities has made it an easy process through AI-enabled sensors and other systems. The proportion of usage of water and electricity in various parts of the cities is conveyed to the concerned authorities and this helps in the optimal utilization of resources, optimal energy consumption and eventual cost efficiency. Smart streetlights are also an example of optimal usage of energy consumption. ML models are helpful in analysing chemical emissions from the vehicles and alert the authorities to take measures in this regard so that environmental protection can be ensured.

Although there are many positive aspects in the usage of AI in BRICS economies, there are certain challenges as well which are faced by the smart cities. Many city authorities are facing problems regarding the installation of hardware and the maintenance costs associated. High technology-enabled systems and the usage of AI-backed technologies require a huge budget for financing high-end projects. The equipment that is installed needs servicing also and for regular maintenance, recurring costs are a major concern. Sometimes natural disasters also damage the AI equipment, which increases the repair and maintenance cost. Privacy concerns are an issue along with societal gaps in terms of usage of AI-based systems and services. A major issue in BRICS can be observed in the digital divide of the nations. Not only is the urban population divided digitally, the gap between rural and urban is also increasing at a very fast pace.

The contributing factors to this digital divide are lack of training in AI, data management problems, lower access to digital gadgets, illiteracy in a few cases, poverty and many other factors. These issues can be resolved only when the governments, the authorities and the stakeholders of the BRICS nationals will develop their capability of adopting AI and ML by bringing adequate reforms, restructuring the related policies and ensuring adequate participation by the citizens by giving them proper resources and training. The inclusivity of the citizens in the AI-based smart cities will not only play a requisite part in the evolution of smart cities in an effective manner but also contribute towards the growth of BRICS economies as a whole.

ABBREVIATIONS

AI	Artificial Intelligence
AMRUT	Atal Mission for Rejuvenation and Urban Transformation
ANPR	Automatic number plate recognition
ATM	Automated teller machine
AVLS	Automated vehicle location system
BRICS	Brazil, Russia, India, China and South Africa
C4AI	Centre for Artificial Intelligence
CCTV	Closed circuit television

COR	City operation centre
COVID	Coronavirus disease
DMA	Direct memory access
DR	Disaster recovery
EU	European Union
GPRS	General packet radio service
GPS	Global positioning system
HRIDAY	Heritage City Development and Augmentation Yojana
IBM	International Business Machines
ICRISAT	International Crops Research Institute for the Semi-Arid Tropics
ICT	Information communication and technology
INR	Indian National Rupee
IoT	Internet of Things
IP	Intellectual property
ISCDL	Indore Smart City Development Limited
ITES	Information Technology Enabled Services
ITM	Information Technology Management
LSCL	Ludhiana Smart City Limited
MAC	Media access control
ML	Machine learning
NASSCOM	National Association of Software & Service Companies
NBBJ	Naramore, Bain, Brady & Johanson
NGO	Non-Government Organization
PCI	Per capita income
PIS	Personnel Information System
PwC	Price Water House Coopers
R&D	Research & development
RFID	Radio frequency identification
SBM	Swacch Bharat Mission
SCADA	Supervisory control and data acquisition
SCADL	Smart City Ahmedabad Development Limited
SDG	Sustainable Development Goal
SE4ALL	Sustainable energy for all
SIDCOP	Sino-Indian Digital Cooperation Platform
SMAC	Social, mobility, analytics & cloud
SMS	Short message service
STEM	Science, technology, engineering & mathematics
STI	Science technology & innovation
STP	Signal transfer point
SWM	Solid waste management
TMS	Transport management system
UK	United Kingdom
UMIAS	Unified Medical Information and Analytical System
UNCTAD	United Nations Conference on Trade & Development
UP	University of Pretoria
USA	United States of America

REFERENCES

Abaker, I., Hashem, T., Chang, V., & Anuar, N. B. (2016). The role of big data in smart city. *International Journal of Information Management*, *36*, 748–758.

Aguilera, U., Peña, O., Belmonte, O., & López-de-Ipiña, D. (2017). Citizen-centric data services for smarter cities. *Future Generation Computer Systems*, *76*, 234–247. https://doi.org/10.1016/j.future.2016.10.031.

Allam, Z. (2017a). Building a conceptual framework for smarting an existing city in Mauritius: The case of Port Louis. *Journal of Bio Urbanism*, *4*, 103–121.

Allam, Z. (2017b). A theoretical application of the extended metabolism model in Port Louis in a bid to promote urban sustainability. In *2nd international conference on energy, environment and climate change (ICEECE 2017)*, Mauritius. (pp. 232–334).

Allam, Z. (2018a). Contextualizing the smart city for sustainability and inclusivity. *New Design Ideas*, *2*, 124–127.

Allam, M. Z. (2018b). Redefining the smart city: Culture, metabolism and governance. In *Case study of Port Louis*. Mauritius: Curtin University.

Al-Kodmany, K. (2018). The sustainability of tall building developments: A conceptual framework. *Buildings*, *8*, 2–31.

Bari, A. (2017). *Working with big data: Scaling data discovery*. Westmount, Canada: Abdallah Bari.

Barton, D., Woetzel, J., Seong, J., & Tian, Q. (2017). Artificial intelligence: Implications for China.

Batty, M. (2016). Big data and the city. *Built Environment*, *42*, 321–337.

Barkham, R., Bokhari, S., & Saiz, A. (2018). *Urban big data: City management and real estate markets*. MIT Center for Real Estate and DUSP January.

Bhadani, A. J. (2016). Big data: Challenges, opportunities and realities. In M. K. Singh, & D. G. Kumar (Eds.). *Effective big data management and opportunities for implementation* (pp. 1–24). Pennsylvania: IGI Global.

Brutti, A., De Sabbata, P., Frascella, A., Gessa, N., Ianniello, R., Novelli, C., Pizzuti, S., & Ponti, G. (2019). Smart city platform specification: A modular approach to achieve interoperability in smart cities. In *The Internet of things for smart urban ecosystems* (pp. 25–50). Berlin/Heidelberg, Germany: Springer.

Caprotti, F., Cowlet, R., Datta, A., Broto, V. C., Gao, E., Gorgeson, L., Herrick, C., Odendaal, N., & Joss, S (2017). The new urban agenda: Key opportunities and challenges for policy and practice. *Urban Research and Practice*, *10*(3), 367–378.

Davenport, T., & Ronanki, R. (2018). Artificial intelligence for the real world. *Harvard Business Review*.

Davey, B., & Tatnall, A. (2015). The internet of things and beyond: Rise of the non-human actors. *International Journal of Actor-Network Theory and Technological Innovation (IJANTTI)*, *7*, 56–67.

Castro, D., & New, J. (2016). The promise of artificial intelligence, Centre for Data Innovation 3.

Fan, J., Han, F., & Liu, H. (2014). Challenges of big data analysis. *National Science Review*, *1*, 293–314.

Greene, G. 2018. Potholes, rats and criminals: A framework for AI ethical risk, Data Smart city solution. Cambridge, MA: Harvard Kennedy School.

Heeks, R., & Shekhar, S. (2018). *An applied data justice framework: Analysing datafication and marginalized communities in cities of the global south* [Working Paper 7]. University of Manchester: Manchester.

Hildebrandt, M., & Gutwirth, S. (2008). *Profiling the European citizen: Cross-disciplinary perspectives*. Dordrecht: Springer Netherlands.

Kitchin, R. (2014). 'The real-time city?' Big data and smart urbanism. *GeoJournal*, *79*, 1–14.

Kummitha, R. K. R., & Crutzen, N. (2017). How do we understand smart cities? An evolutionary perspective. *Cities*, *67*, 43–62.

Markotkin, N., & Chernenko, E. (2020). *Development of artificial intelligence technologies in Russia: Objectives and reality.* Carnegie Moscow Center.

Mazzara, M., Zhdanov, P., Bahrami, M. R., Aslam, H., Imam, M., & Salem, H. (2021). Education after COVID-19. In *Smart and sustainable technology for resilient cities and communities – Advances in sustainability science and technology.* Berlin, Heidelberg: Springer.

Mingaleva, Z., Vukovic, N., Volkova, I., & Salimova, T. (2020). Waste management in green and smart cities: A case study of Russia. *Sustainability, 12*, 94.

Muzammal, M., Qu, Q., & Nasrulin, B. (2019). Renovating blockchain with distributed databases: An open source system. *Future Generation Computer Systems, 90*, 105–117.

Neirotti, P., Marco, A. De, Cagliano, A.C., Mangano, G., & Scorrano, F. (2014). Current trends in Smart City initiatives: Some stylised facts. *Cities, 38*, 25–36. https://doi.org/10.1016/j.cities.2013.12.010.

Ni Loideain, N. (2017). *Cape Town as a smart and safe city: Implications for governance and data privacy.* International Data Privacy Law.

Raghuvanshi, R., & Sharma, T. (2023). Smart cities in emerging economies: Opportunities, challenges and policy implications. In Goyal, D., Kumar, A., Piuri, V., & Paprzycki, M. (Eds.). *Proceedings of the third international conference on information management and machine intelligence. Algorithms for intelligent systems.* Singapore: Springer. https://doi.org/10.1007/978-981-19-2065-3_36

Roberts, H., Cowls, J., Morley, J. et al. (2021). The Chinese approach to artificial intelligence: an analysis of policy, ethics, and regulation. *AI & Society 36*, 59–77.

Santanen, E. (2019). The value of protecting privacy. *Business Horizons, 62*(1), 5–14.

Siew, G., & Allam, Z. (2017). Culture as a driver for sustainable urban development. *UIA 2017.*

Souza, Weslei Gomes de, Pereira de Melo, E.R., De Souza Bermejo, P.H., Sousa Farias, R.A., & Gomes, A.O. (2019). How and where is artificial intelligence in the public sector going? A literature review and research agenda. *Government Information Quarterly, 36*(4), 101392. ISSN 0740-624X, https://doi.org/10.1016/j.giq.2019.07.004.

Tole, A. A. (2013). Big data challenges. *Database Systems Journal, 4*, 31–40.

Vogt, J., & Portmann, E. (2017). A proposal for the next generation of courier delivery and storage in the shared economy. *Wirtschaftsinformatik in Theorie und Praxis: Festschrift zu Ehren von Prof. Dr. Andreas Meier*, 119–128.

Voronezh,SmartTrafficLights. (2021).

Voto, A. (2017). Blockchains and the civic nervous system. *The Journal of Field Actions, 17*, 60–63.

Zimet, D. M., & Jacob, T. (2001). Influences of marital conflict on child adjustment: Review of theory and research. *Clinical Child and Family Psychology Review, 4*, 319–335.

FURTHER READING

Website Links

Artificial Intelligence: An Opportunity for Growth, Development, and Democratization. European Parliament. 2019.

City of Cape Town Open Data Portal. https://odp-cctegis.opendata.arcgis.com/

City of Cape Town. (2018). CCTV adds another layer to ShotSpotter technology.

Clarke, A., and & Margetts, H. 2014. Governments and Citizens Getting to Know each other? Open, Closed, and Big Data in Public Management Reform. Policy and Internet.

Commentary on the implications of AI on copyright globally and in South Africa, especially concerning proposed amendments to the current Copyright Act. For example: Rens, A. 2019. Open Data South Africa Toolkit. Contracting and licensing of data service providers OECD. 2019. Recommendation of the Council on Artificial Intelligence.

Data Science for Impact and Decision Enablement, Council for Scientific and Industrial Research. GCRO https://www.gcro.ac.za/

Department of Cooperative Governance and Traditional Affairs (COGTA). 2016.

Department of Telecommunications and Postal Services (DTPS). 2017. Digital Society.

Ernst & Young, FIAC survey: foreign business in Russia (2020)

Five-Year Integrated Development Plan: July 2017 – June 2022.

For a South African take on 'smart cities from the bottom up', see Odendaal, N.

http://refhub.elsevier.com/S2210-6707(17)31112-5/sbref0535

http://refhub.elsevier.com/S2210-6707(17)31112-5/sbref0685

http://refhub.elsevier.com/S2210-6707(17)31112-5/sbref0720

https://data-mining.philippe-fournier-viger.com/china-international-bigdata-industry-expo-2019-a-brief-report/

https://doi.org/10.1007/s00146-020-00992-2

https://economictimes.indiatimes.com/topic/BRICS-Smart-Cities-Conference/news

https://indianexpress.com/article/india/smart-city-awards-surat-tops-list-of-cities-using-data-for-delivering-governance-7376339. https://www.chinahighlights.com/travelguide/top-china-smart-cities.htm

https://smartcity.press/brazil-smart-cities/

https://thechinaguys.com/china-smart-cities-development/

https://unctad.org/news/few-developing-countries-overperform-frontier-technologies-most-lag-behind

https://unctad.org/page/technology-and-innovation-report-2021

https://www.financialexpress.com/industry/nasscom-launches-one-more-digital-plaza-in-china-to-provide-market-access-to-indian-it-firms/1182782/

https://www.prnewswire.com/news-releases/chengdu-making-strong-push-to-foster-next-gen-ai-technology-301181080.html

https://www.tmforum.org/press-and-news/can-south-africa-deliver-on-its-smart-city-dreams/

Integrated Development Plan 2017/8 – 2021/2, Second Edition.

Integrated Urban Development Framework.

International Journal of Information Management Ministry of economic development of Russia, Deputy Minister Oksana Tarasenko in an interview with CNews - about the development of artificial intelligence in Russia (2021).

International Journal of Information Management Oxford Insights, Government AI Readiness Index 2020.

Ministry of Housing and Urban Affairs (MoHUA), "Vision".

Ministry of Urban Development, Smart City: Mission Statement & Guidelines, New Delhi: Government of India, June 2015.

MoHUA, SCM Monthly Bulletin, Issue # 3, June 1, 2021.

MoHUA, SCM Weekly Bulletin, Issue # 18, June 5, 2020.

MoHUA, SCM Weekly Bulletin, Issue # 21, June 26, 2020.

MoHUA, "What do we mean by a Smart City".

National Development Plan 2030: Our Future, Make it Work. Pretoria: Republic of South Africa.

National Integrated ICT Policy White Paper. 2016.

National Planning Commission. 2012.

Nelson Mandela Bay Municipality. 2017.

Presidium of the Government Commission on Digital Development, the Use of Information Technologies to Improve the Quality of Life and Business Conditions, Passport of the federal project "Digital Technologies" (Protocol № 9 of 28.05.2019).

Privacy International. 2019. State of Privacy: South Africa. https://privacyinternational.org/state-privacy/1010/state-privacy-south-africa#dataprotection

Project Matla: Energy Analysis for the City of Tshwane.

Smart city: Neoliberal discourse or urban development tool? In Greugel, J., and & Hammett, D. (eds). https://link.springer.com/chapter/10.1057/978-1-137-42724-3_34

South Africa: South Africa's National E-Strategy – Towards a Thriving Future.

TASS Agency, Putin called artificial intelligence the basis for a new breakthrough in the development of mankind (2020).

TASS Agency, The federal budget deficit of Russia in 2021 will amount to 2.75 trillion rubles.

The Constitution of the Republic of South Africa Act 108 of 1996.

The Palgrave Handbook of International Development. London: Palgrave MacMillan.

Tomer, A. (2019). Artificial intelligence in America's digital city. Report: Brookings.

UN-Habitat. (2016). New Urban Agenda. http://habitat3.org/wp-content/uploads/NUA-English.pdf

Understanding algorithmic decision-making: Opportunities and Challenges, Report: Panel for the Study of Science and Technology. 13.

11 Strategies for the Sustainable Development of the Smart Cities by Implementing AI

A. Stephen Sagayaraj[1], *S.K. Kabilesh*[2], *T. Kalavathi Devi*[3], *and Varun Chhabra*[4]
[1]Bannari Amman Institute of Technology, Erode, India
[2]Jai Shriram Engineering College, Tirupur, India
[3]Kongu Engineering College, Erode, India
[4]Quantum Age Solutions Ltd., Hillingdon, United Kingdom

11.1 INTRODUCTION

According to the survey of the United Nation's Department of Economic and Social Affairs, 55% of the world's population migrated to cities and it will increase to 68% by the year 2050. This migration will lead to increase in population density, excessive use of resources, need of better infrastructure, and improved quality of life [1]. The people's migration from the rural areas to cities, conversion of rural areas into urban, implementing technologies in transportation, building new infrastructure, and many others are the real problems in day-to-day life [2]. The rising pollution levels caused by the increasing number of technologies create problems for human beings and other lives [3]. The reality that exists in the atmosphere is the emissions of greenhouse gases like CO_2, methane, and nitrous oxide as well as the presence of natural resources. Even COVID-19 has a significant impact, which has been lessened by ongoing lockdown implementation, which results in a halt in economic growth. Therefore, some adjustments to urban growth are required to create future habitats. [4]. The sustainable development is an immediate action to take place from public and private sectors [5]. Learning about sustainable development and putting the idea of smart cities into practice create the conditions for a sustainable urbanism [6]. The technological growth and transformation are to be done by funding the smart city activities [7, 8].

Artificial intelligence (AI) provides an undetermined development to the society to bring comforts to the day-to-day life of the people living in the urban areas [9, 10]. An important means of implementing smart culture in urban areas is made possible by the development by delving deeper into the digital idea. By developing technology ideas, humans could have health problems, which should also be taken into account. [11–14]. The dynamic capability theory is a framework to obtain a highly dynamic environment to provide advantages to a business environment [15, 16]. The complexity

DOI: 10.1201/9781003225317-11

theory provides an essential way to enable organizations to analyze complex problems [17, 18]. This theory can identify the internal and external factors in an organizational sector [19]. The strategic developments of urban cities are made by identifying key factors such as time, security, efficiency of energy, and reduction of pollution [20–22].

The research of sustainable development in smart cities is classified into various sectors. In the very first section, we establish the framework of cities to determine research fundamentals. Moreover, the research methodology of smart cities' development using AI is carried out in the second section. Next, real-time applications of smart cities using AI are discussed. In the final section, limitations and future scope are discussed.

11.2 THEORETICAL BACKGROUND

The AI turns into the high-level procedures that are implemented in everyday life [23, 24]. It is fundamentally a high-level science innovation that is used in each application region [25, 26]. It builds regions, in which research studies are going on in more significant ways [27]. It demonstrates a self-acting machine that produces effective results [28].

The present AI is a combination of AI and profound learning strategies, which establish the most recent forward leap in AI [29, 30]. Association learning is key to capacity to exploit AI [17]. Being able to innovate is also crucial to the mechanical component of AI. Additionally, AI can boost associations' development and growth. As [31], advancement capacity is likewise significant. Given the ecological dynamism and the present review's emphasis on the abilities of AI to develop smart cities, we can characterize the idea of dynamic capacity. The phrase was defined by Teece et al. [15] as 'the company's capacity to incorporate, form, and reconfigure inner and outside skills to address quickly evolving conditions', which brought about the dynamic capability theory.

We have considered these kinds of hypotheses so that our review can work with AI to advance learning capacity, development ability, and innovation capacity that develop in a dynamic, evolving climate. We try to assess the degree of the progressions that would happen in a city that initiates new improvement exercises, the quantity of new cycles presented, the accentuation on creating advances or potentially mechanical development, and the regional administration's accentuation on R&D, mechanical authority, and advancements [32, 33]. We subsequently investigated AI as an inventive type of business venture dependent on the association's innovation. Eventually, the authoritative capacities investigated in this review are dynamic abilities which give an upper hand in light of the fact that the urban communities are additionally evolving. Simulated intelligence inspires these abilities generally in learning in fields of development and innovation [15, 16, 31].

11.2.1 Complexity Theory

We would like to only analyze AI uses internationally to see how significant it is for smart societies and organizations [34]. Facebook is of late uncovered designs to put resources into AI at the yearly F8 gathering. Baiduis put it intensely in computerized reasoning: 'The time of versatile Internet is finished'. Indeed, 5G, which centers on AI, is turning into a problematic advancement of the Internet. JPMorgan uses AI to accomplish high monetary mechanization and increase representative usefulness.

Shelby Austin, CEO of Deloitte, stated that the association is eager to collaborate with the World Economic Forum and the public authority of the United Kingdom on improving acquisition rules that help the public sector bring AI as the assistance of its elector in a way that is both productive and moral.

Moreover, all the biggest organizations throughout the planet use AI to work on their associations and embrace critical activities to upgrade their presentation. These activities include purchasing media, automating run books, reducing the amount of creation the board is working on, resolving customer development issues, identifying and blocking security intrusions, fitting advancements (online or detached), checking on web media comments to determine overall brand inclination and issues, anticipating future customer purchases and making offers as necessary, and measuring internal consistency by means of internal audits. [35]. According to an intricacy point of view, AI and development are essentially identified with the complexity theory. As they empower an association's inventiveness through powerful organization structures [19], the new imaginative information (AI) obtained by associations [36] makes collaborations that can decide the advancement of associations and empower imaginativeness [19]. In addition, the more noteworthy the availability of society and associations, the more data available, and the more perplexing innovative frameworks are. As a result, organizations create new, challenging, pioneering biological systems using a variety of data-gathering techniques.. Intricacy hypothesis is useful for the heterogeneous specialists for advancing networks. The applications of AI can make people learn the significance of AI in a region or a city, which can assist a city with accomplishing its objectives of turning into a smart city.

11.3 SUSTAINABILITY AS AN IMPORTANT NEED

Computer-based intelligence applications can anticipate the execution of time by the achievement that would depend on progressive data and given modified objects, improvement cycles, or administrations for web-based search and web-based media clients. The University of Queensland fostered a profound learning framework called Blackboard [37], a neural organization's examination of player's dynamic cycles assessed on their past conduct, to assist players with settling on the best choices to achieve their objectives.

This element of AI would affect regional authorities. For instance, it would dissect drivers' ebb and flow circumstances in urban communities and make expectations concerning what will occur straightaway to stay away from car crashes or gridlocks, manage surveys, and so forth. This component would probably be used in instruction, to analyze residents' behavior to identify which of their tendencies are harmful, correct them, or even to discover the problems that residents face and encourage the development of better designs. Computer-based intelligence subsequently empowers manageability as a genuinely necessary change in urban communities.

11.4 PROBLEMS FACED

This segment presents four significant issues at the worldwide level in the current enormous urban communities, in view of the different sources accessible [38–40]. It then, at this point, proposes key activities to embrace accordingly.

11.4.1 Migration of People

This review is roused due to the disturbing information from different authority sources. As per the current information on the United Nations (UN), the population density increase of metropolitan collection has expanded at a confounding rate. The total population is relied upon to increase by more than 2000 million in the following 30 years, arriving at 11,000 million by 2100 at its extreme basic point. Population density increase and conveyance (increase in the population living in metropolitan agglomerations) have less-effective results for the climate and the sustainable development goals (SDGs), settled among, to diminish ecological effect and work on friendly prosperity. Ongoing investigations of the expansion in the population rate in the nine urban communities with the most noteworthy outstanding development in population density overall rate say that India will supplant China as the extreme crowded country on the planet by 2027 [41].

The United Nations Educational, Scientific and Cultural Organization (UNESCO) [42] gauges that, by 2030, more than 60% of the total population will occupy metropolitan focuses. As indicated by the UN [1], by 2050, there will be more than 41 big cities (of more than 10 million occupants). The pattern toward metropolitan population expands in the worldwide population (Figure 11.1), and a higher level of the metropolitan population occupies metropolitan agglomerations (right now, almost half).

In Spain, around 60% of the people now live in metropolitan areas. Beginning around 1950, the population density has increased around 53% in several cities, whereas the urban population has grown by almost 136%. Such development creates a narrow provincial differentiation, which makes it hard to keep up with practical advancement [43].

Further, most created nations have transformed population pyramids because of an extravagance and luxury anticipation upon entering increasing population, whereas the less-developed nations have higher success rates. Endeavors to work on the individual satisfaction are higher in evolved nations as those quite often

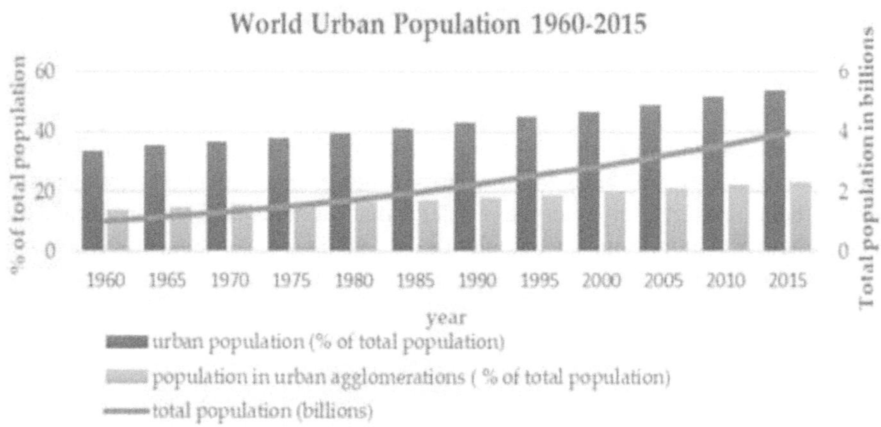

FIGURE 11.1 World population evolution.

diminish disparities and battle craving and unhealthiness. Rather than countries like Africa and other creating districts, created nations have minimum rates of birth and expanded resettlement. Movement hence turns into a trigger of overpopulation in many cities around the world. The dramatic expansion in transient development impacts the climate, delivering one more serious issue of 21st-century huge urban areas: over-the-top contamination.

11.4.2 Growth of Metropolitan Areas

Today, individuals are genuinely keen on medical problems and the ramifications of the climate for their well-being. Ecological well-being arose as an issue during the last century because of dramatic modern turns of events. Different distributed investigations show the repercussions of this turn of events: one out of every four infections is caused by contamination [8].

Highlights like these alert societies. Notwithstanding the doubt of many, this news propels a perplexing change in the example through the mission for game plans. The level of particulate matter air tainting (PM 2.5) turns out to be general basically [39]. Figure 11.2 shows the extension in the ordinary yearly receptiveness to air tainting levels of PM 2.5 in micrograms per m^3, and in complete level of the general population introduced to levels over the cutoff focuses set up by the World Health Organization (WHO) [40]. Of the years for which data are open, we see that 2016 shows the most essential recorded tainting indisputably similar with the level of the general population revealed.

The current metropolitan peculiarity started with the Industrial Revolution and has been creating in various metropolitan models, with changes in friendly examples and new ways of life because of development and the appearance of data advancements (IT). As Figure 11.3 shows, the correlation of a city like Gwalior (India) (one of the most dirtied urban communities on the planet) to the reference information set up by the WHO shows that Gwalior far surpasses the breaking point esteems set

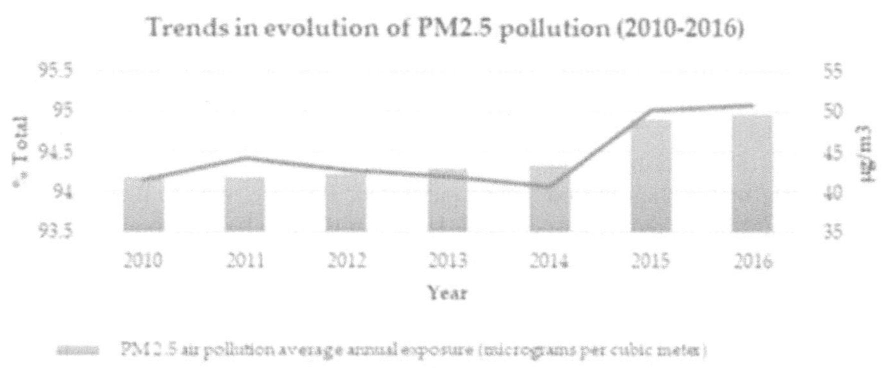

FIGURE 11.2 Comparison of pollution exposure.

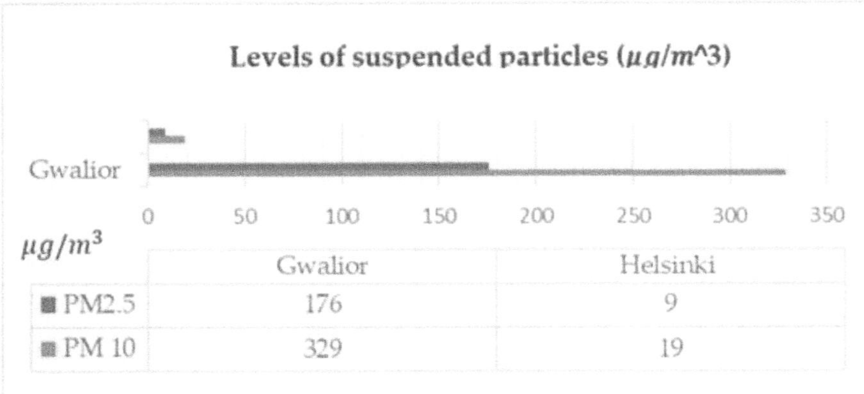

FIGURE 11.3 Levels of particle suspension in Gwalior and Helsinki.

up [45]. The world's cleanest urban communities, such as Helsinki (Finland), conversely, have values beneath as far as possible [9].

The information confirms the critical requirement for activity in the most dirtied urban areas. This degree of air contamination escalates environmental change through ecological catastrophes, for example, corrosive downpour and the decay of the ozone layer.

11.4.3 The Problem in Environmental Prosperity

Disregarding the fact that there is still no settlement on the advancement of SARS-CoV-2 (COVID-19), it is by and large acknowledged that tainting and overpopulation have obliged animals to live closer to metropolitan regions, extending the probability of scenes of new ailments, for instance, COVID-19. But if we don't start using our resources effectively, more illnesses will spread throughout the world, posing new problems for humanity. Given this issue of pollution and the constant COVID-19 pandemic, a basic point of convergence of investigation is prosperity and the effect of the environment on prosperity. Environmental prosperity is one of the focuses that have mixed the most interest of late, following the exceptional current progression of this earlier century. One of every four sicknesses is associated with defilement factors [44]. The tainting regards at present that are correct now being made are significantly higher than the midpoints permitted. Because of SO_2, they can even show up at a value on numerous occasions higher in 10 min all throughout the planet. In case we continue at such extraordinarily evident levels, more afflictions could emerge, and current sicknesses could fall apart. Metropolitan people group, such as Helsinki (Finland), get values far underneath the typical recognized by the WHO and thus become worldwide models. In any case, the overall data show the desperate need to act in each of the most tainting metropolitan networks throughout the planet, as air defilement affects ecological change, ordinary fiascoes (e.g., destructive storm), and breaking down of the ozone layer. Greater Expenditure on Energy

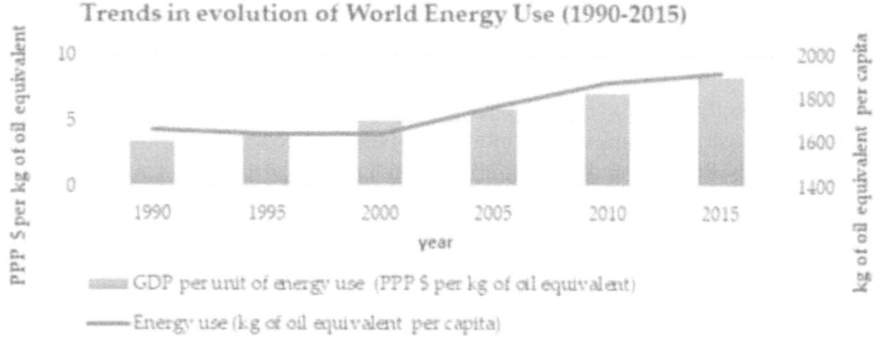

FIGURE 11.4 Usage of energy.

Consumption (Almost 80% of World Consumption) always increases even the available resources are decreasing in the real world as the usage of oil increases [45], even the oil reserves deploying continuously is shown in Figure 11.4.

11.5 HOW SMART CITIES ARE USING ARTIFICIAL INTELLIGENCE

There are various advances where AI can assume a role. From the beginning, AI permits separating geospatial highlights to be used in guides and GIS programming in a robotized way. Notwithstanding, it is great to make reference to different regions that come after this first phase of extraction, such as the investigation of spatial-fleeting metropolitan examples (metropolitan development, foundation sending, portability and neighborhood relations, and so forth) Lastly, with geospatial designs advanced with quality metadata, for instance, data on building use, transportation schedules, and so forth can be fed to AI&ML to examine patterns and heterogeneities present in a city.

AI is now used in the plan and framework of the board of urban areas. It will be an irreplaceable instrument in an ever-increasing number of explicit situations as we push ahead; however, here are only a couple of models. Mobility continues to be a focus that considers self-driving cars. Independent innovation is an expansive and complex innovation region all by itself, yet the item as of now incorporates AI. As the administrative and consistency highlights are fulfilled for well-being, this will turn into a huge element in how city travel courses are planned and overseen. Spacemaker AI is an organization in Oslo that has become eminent in the course of the last year. They use AI innovation to expand the capability of building locales for a given arrangement of rules, such as perspectives, daylight, commotion, water spillover, and access. In a real sense, billions of various variables can be evaluated rapidly and used to plan the structural impression of enormous scope improvements. Sewage systems in urban areas become more effective when using artificial intelligence to predict the amount of synthetics that should be used to clean up the water spillover, given different potential situations, such as preparing for regular population growth and precipitation levels.. With IoT (Internet of Things), AI will turn into the standard way in which we plan most things.

11.5.1 The Solutions for Implementing AI

Although there are hazards inborn in conveying novel advancements, such as AI, the benefits will imply that the developers of 'smart city' innovations will need to see how to address those dangers so they can receive the rewards of having the option to more readily use and comprehend their information. According to Standard GDPR (General Data Protection Regulation), which entails good data governance and implementing privacy by design and by default is also apply to associations, such as public and nearby states and citizens of frameworks.

To deal with the support of these basic resources, we want to get monetary evaluations for progressing upkeep. It is significant to build up an interaction that distinguishes harm as ahead of schedule as could be expected. Recognizing surface harm and signs, such as breaking, potholes, subsidence, or disintegration, are frequently the main pointers that pressing healing activity is required.

Therefore, the assets expected to evaluate street conditions are generally exceptionally high, requiring impressive measures of work (manual looking over for the most part) and complex undertaking intending to stay away from the misdesignation of crucial assets and inefficient financial plan designation. Among different kinds of resources that are normal in the urban communities, there are structural utility designs, for example, power transformers, arches and microwave pinnacles or telephone posts, rails, gasometers, and siphoning stations.

The reasons for this undertaking have been:

to permit the city gathering to oversee street upkeep costs preemptively,
to ideally map public asset use needs for redesign or development, and
to give a technique to all future public foundation support endeavors.

11.6 HOW CAN ARTIFICIAL INTELLIGENCE FACILITATE THE DEVELOPMENT OF SMART CITIES?

The smart city headways needs more data which is portrayed as high volume, fast, or as high variety information resources, which includes the various data sources with assorted datasets.

Huge data and man-made thinking (AI) are interlinked. Man-made insight implies various methods 'for using a non-human structure to acquire indeed and imitate human shrewd conduct'. AI can capably channel through tremendous measures of Big Data to make data conjectures and wise responses to fuel smart city advancements.

The way wherein this works depends upon whether the AI is regulated or unaided. In directed learning, datasets and target regards are made to plan AI associations to find unequivocal courses of action in the assembled rough data.

The AI will then complete tweaked tasks and exercises, while examining new opportunities and potential results that may give favored outcomes over current game plans.. In solo learning, datasets that are unmarked and unassembled are used for posture requests of AI associations, which will subsequently search the data for inactive characteristics and hidden examples.

11.7 REAL-TIME IMPLEMENTATION OF AI IN SMART CITIES

11.7.1 Improve Road Infrastructure with AI-Based Traffic Systems

Limited streets in urban communities can be an enormous bother. The blend of little back roads and substantial trucks, in some cases driven by laborers who are new to the area, has brought about endless examples of huge vehicles running off the street, turning over on thin turns, or being engaged in mishaps. Issues, such as traffic and blockages, are additionally normal, particularly in metropolitan communities.

The undeniable answer to settle this is to update the current street foundation or create new roadways; however, precisely anticipating their need and upkeep timetable can be troublesome. This can be fixed with the assistance of AI-based traffic and versatility frameworks. These frameworks gather and investigate traffic information, create arrangements, and apply them to traffic foundations. AI-based traffic frameworks work by gathering information from associated traffic frameworks, which gives input about live traffic or long stretches of authentic traffic conduct. These frameworks can likewise consider information about a large group of other potential issues that can affect streets and traffic later on. These frameworks use AI models to process, analyze, and discover information about traffic foundation. The AI then, at that point, uses those experiences to propose arrangements that can empower specialists to assemble a better and more secure framework. A deep learning calculation can likewise help in effectively foreseeing upkeep necessities of street frameworks and along these lines forestall costly and impromptu disappointments.

A Cambridge-based AI organization, Intelligence, reported subtleties of a proof-of-idea project with worldwide development and framework bunch Skanska, which could save Hampshire County Council in overabundance of £100,000 per annum on street seepage and crevasse upkeep work. In a preliminary, seepage information of Hampshire County went through a profound learning arrangement that transformed it into a progression of models that could empower partners to foresee where waste issues were probably going to happen.

One more model is the programmed acknowledgment of street harm by ARCADIS, in which pictures are taken with a camera of a specific street segment using yearly pictures from CycloMedia or with a GoPro. These pictures are used to decide the specific area of the deformity, sort and degree of imperfection. Using chronicled information, this model can likewise ascertain the decay of streets, permitting street specialists to draw up a prescient support plan.

11.7.2 Water Resource Management

The immense organizations of covered water, wastewater, and stormwater foundations are the veins and corridors taking care of our kin, urban communities, and ensuring our current circumstances. In the absence of feasible water, squander, and stormwater arrangements, the personal satisfaction cannot be met out. Society's water, wastewater, and stormwater frameworks consistently play critical roles in wiping out infection, shielding the climate, and securing networks.

There is a huge strain on water utilities capacity to keep conveying basic administrations to clients, while working with restricted assets. However, by using AI in water assets, executives can clear a path for need-based water supply and water-saving frameworks.

Utilities can send shrewd meters, sensors, and other IoT equipment that will unavoidably bring about the assortment of gigantic measures of information about water consumption of an area and can likewise distinguish spillages to suggest speedy activity. Mechanized examination of this information using machine learning and AI will empower building a productive water framework, upgrading ebb and flow water assets, foundation anticipating water designs, and give fundamental adaptability to deal with unexpected conditions.

11.7.3 Air Quality Monitoring

Different drives have made organizations of sensors equipped for catching air tests, dissecting them, and educating the overall population regarding the air quality in both huge and little urban communities. Notwithstanding contamination signs, the estimation of petroleum-derivative particles is not really set in stone, despite contamination indicators, the assessment of petroleum-derived particles is not necessarily fixed.

11.7.4 AI for Future Proof Infrastructure

The problem of urban areas is to understand the increase of the energy transition, such as restricted space accessibility, scant workers limit and materials, and restricted monetary assets. To accomplish the environmental objectives and speed up the energy change, mass remodel and development works will be inevitable. Computer-based intelligence empowers nearby legislatures, development organizations, service organizations, and different partners to handle these difficulties. For example, AI for spatial item acknowledgment, dependent on satellite symbolism, blended with AI for course advancement, can streamline framework arranging in the restricted space and support in focusing on work and adjusting that to the restricted work and materials. Accomplishing the environmental objectives likewise means that urban communities need to use all accessible nearby wellsprings of practical energy, making sustainable energy frameworks with decentralized environment-friendly power sources, affecting the framework with their discontinuous character.. Innovative fifth-generation area warming and cooling lattices is to include heat furthermore, power as energy transporters toward environment nonpartisan hotness, and power arrangement in the gathered climate. Furthermore, these lattices incorporate the metropolitan water networks as an identical wellspring of low-temperature heat, making incorporated multi-item networks in our urban areas. To help conveyance framework administrators and energy suppliers, AI application can support understanding the interlaced energy networks and support in the expense-proficient control of these frameworks. ...In AI-powered network, the function includes the inexhaustible energy age such as the sun, wind, geothermal and water nuclear power for the further usage and for the network control.

AI-driven control of consumer energy usage also empowers request side management for network adaptability, it can be applied on the customer side to anticipate further improved energy effectiveness. These applications make the network stronger for decentralizing further, questionable age, empowering the environment-friendly power, whereas residents are upheld in taking on adaptable and effective energy-use designs, which helps the families and the network. Models are the Netherlands-based Quby, a piece of Eneco, fostering these administrations for service organizations in the Netherlands, Belgium and Spain, and Germany-based Tado7.

11.8 RENOVATION WAVE STRATEGY IN EUROPE

In 2020, the main goal was energy gains and monetary development;, the Commission distributed another procedure to help the redesign which is 'A Renovation Wave for Europe," which means Greening our structures, making occupations, further developing lives.

This technique expects to fold yearly energy remodel rates in the following ten years. Just as decreasing outflows, these remodels will improve personal satisfaction for residing individuals and who using the structures and also ought to make numerous extra green positions in the development sector. Almost 34 million Europeans incapable of keeping their homes warm appropriately, and redesigning additionally handles energy need. It can address the well-being and prosperity of weak individuals, while lessening their energy bills – as illustrated in the Commission proposal on energy destitution, likewise a piece of remodel wave methodology.

The Commission has taken on new standards for the 'shrewd availability of structures'. Another brilliant status marker plans to advance computerized cordial redesign, incorporate environmentally friendly power, and empower estimation of genuine energy consumption. At a similar time, the Commission dispatched the New European Bauhaus and it gives a discussion where Europeans can meet up to share thoughts on environment -disposed design. It contains three stages: co-plan, conveyance, and scattering. The plan stage was dispatched in January 2021 and will prompt a progression of EU-supported tasks in near future to rejuvenate plan thoughts in about five unique areas across the EU.

The remodel wave drive will expand on measures concurred under the 'Perfect energy for all Europeans' bundle, quite necessary for every EU nation to distribute a drawn-out building redesign methodology, different parts of the Directive on Energy Performance of Buildings, and building-related parts of every EU country's public energy and environment plans.

EU Commission's 2020 Drive for Rebuilding and Restoration: Progress and Implementation

- In 2020, the Commission distributed a guide on the drive to disclose the drive to the general population.
- This was trailed by a public counsel. We got reactions from 187 significant partners, which are summed up in a report.
- Progress was declared in a public statement and its going with Q&A.
- The Commission distributed its correspondence technique and activity plan in October 2020, joined by a record on help from the EU financial plan to open interest in rebuilding model.

- This will be talked about by the EU Council, the European Parliament, other EU organizations, common society, and partners, who will give inputs.
- The Commission is currently zeroing in on carrying out the activity plan.

11.9 SMART CITY IMPLEMENTATION IN SINGAPORE

Singapore has a vision to make the first smart city in the world by furnishing advancement to the individual fulfillment, supporting associations, and building more grounded openings. 'Everyone, Everything, Everywhere, All the time' is the main motive of Singapore smart city implementation. The Smart Nation Platform seems to engage more conspicuous organization, better situational care through data collection, and it is capable of sharing accumulated sensor data.

11.9.1 Vision and Mission

Singapore is in developing aspect of smart cities from 2014 under the program of Shrewd Nation to associate the better living for the people. Distinctive metropolitan networks all through the planet are investigating various roads in regard to the 'insightful city' by using headways to deal with a wide extent of metropolitan to achieve the vision. The vision of the city is a response to creating metropolitan challenges of developing people, metropolitan thickness, and energy practicality. Various accomplices, for instance, development makers and business visionaries all throughout the planet, are free to be locked in with this vision to use of novel musings and splendid courses of action with overall potential. A fundamental occupation of ICT is to start the interpretation of structures in the smart city drives. Achieving such fusion is needed to work with the best usage of resources across different structures.

11.9.2 Transportation and Urban Mobility

In order to set a traffic-free road, Singapore installed the concept of ITS in traffic parts. The backbone of the ITS in Singapore starts from that all-encompassing methodology approach to traffic executives; ITS works together with other vehicle drives such as people's movements in peak hour, a transport quantity framework, well-function city transport frameworks and to upgrade vehicle frameworks in the city. Using ITS parts, it gives a number of savvy mobility functions for residents.

11.9.3 ONE.Motoring

ONE.Motoring is the extensive entry serving helping all the transporters in cities. Traffic details in the street are collected from the video acquired from the roadside cameras and updated in taxis through GPS. Even drivers can see previews that are taken, because safety reasons, continuous moving video or close-up shots are not released. It likewise gives data on flow ERP rates (electrical road pricing), segments where streets closed on working, news, bearing on the road and stopping data. This valuable entrance can likewise be available in cell phones. ONE.Motoring not only gives traffic data yet but, in addition, offers data and direction for residents in regard to purchasing, selling, and keeping up with their vehicles.

The Land Transport Authority (LTA) employs reconnaissance cameras to pay special mind to street episodes. LTA provides assistance to the surveillance team that send the rescue team within 15 minutes to tow the vehicle. 'Your Speed Sign' is an electronic gadget that warns drivers in over-speeding. It urges drivers to remain under the cutoff what's more and, in this manner, further develops security on the streets. The guidance system of parking is introduced was 2008 and this gives drivers on-time data on stopping accessibility. This system reduces the peak time traffic in order to of search parking and advances, which are productive uses of existing stopping facilities. The spacing is shown in the e-board or updated in an online portal either in mobile or through car-accessing devices.

The smart city advances in transportation build the new sophisticated technology for the people. As part of it, self-driving cars and autonomous driving develop to achieve a richer technology in the smart city concept. The MyTransport.SG cell phone application gives continuous data to workers. It refreshes its elements routinely to further develop suburbanites' voyaging experience. Since the latest year LTA has been working with SBS and SMRT together to introduce and test a new framework, which decides continuous transport areas and subsequently gives more precise transport appearance data for more than 4,700 public transports in more than 360 courses. The data stored for maintenance can be accessible by transport companies. Suburbanites are there to identify available spaces to assist the transport of vehicles and for vehicle pickup. Green shows available seats, yellow shows accessible waiting spaces, and red demonstrates canceled standing. Different applications, created by My Transport Singapore, brought to residents by the LTA are accessible to upgrade workers' and drivers' helpful metropolitan versatilities.

11.9.4 Safety and Citizen Surveillance

As per the report by the Economist Insight Unit, Singapore stands ahead as the second most city enriched with surveillance. There is an SMS administration called the Emergency Short Messaging Administration Helpline to put it plainly, which is intended to offer individuals from hard of hearing. The undertaking is upheld by the finance government. The essential plan of this drive is to increase the accessibility to Government e-administrations in accordance with the Government's 2010 master plan. The cops give Internet-based electronic police for residents to advantageously accumulate data, document police report on the web, and handle managerial undertakings, for example, an application for a certified duplicate criminal record. This is to stop crime on EPC and provide an elective Internet-based road for residents to make only few reports or submit data to help the cop in battling wrongdoing.

11.9.5 Emergency and Response

The Singapore Civil Defense Force (SCDF) is a formal association, administered by the Service of Government, which serves for putting out fires, salvaging crises for clinical benefits; assuaging dangerous materials occurrences, just as form; carrying out and upholding guidelines on fire well-being. Crisis in Medical Service (EMS) is undertaken by the SCDF, and it tends to reach through three-digit dial '995'. The 995 Operations Center can also be additionally reached through portable applications that

SCDF give coordinated effort to IDA which intends to expand endurance rate from occurrences, for example, out-of-medical clinic heart failure. The application alleged by my responders has been dispatched as of late in April 2015, which makes clients get aware of adjacent instances of suspected heart capture and directs the users to react before the SCDF as it could be shown up. After the dispatchment of the application, there have been more than 2,000 downloads, despite the fact that the quantity of enrollments to turn into responders didn't arrive at the comparative level until now.

'SCDF' permits residents to give inputs to SCDF, directions for using fire quencher, instructions to do mouth to mouth and AED, just as access to SCDF yearly reports. Different applications that permit people in general with admittance to basic crisis data are 'Organic Threats', 'Purification Methodology', 'Openness to Biological Agents', 'Fire Wellbeing', and 'If An Explosion Occurs'. Many e-administrations are likewise accessible to SCDF's fundamental site. These administrations include as follows: (i) The web installment for SCDF's administrations; (ii) apply for courses at SCDF for government-funded instruction programs; (iii) buy fire and rescue vehicle reports; and (iv) SCDF offices finder. One more significant part of Smart Nation drive that is in transit is telemedication. Singapore points to advance the far-reaching use of wearable advances such as wellness trackers, brilliant watches, and surprisingly shrewd apparel that can screen the prosperity of a patient. The thought is not exclusive to record imperative signs, such as blood pressure, pulse, and internal heat level, which in addition sends the information through Internet to assigned medical services experts or family individuals.

IDA set up its arrangement for Smart Health-Assist pilot project in the Jurong Lake District in late 2015 to help the requirements of the maturing populace of Singapore. It is intended to record information from easy-to-understand sensors in the places of the old and the patients experiencing persistent illnesses to be sent safely online to medical services suppliers, permitting them to screen people, get cautions, and react to any crises.

11.9.6 Environment

As a small country, Singapore has little land to assemble for collecting water. Water insufficiencies will be very hard for Singapore with consistently growing solicitation and as such, Singapore should be impelled to create and encourage limits close by. Under the Service of Environment Water Resources, there are two lawful sheets: the National Climate Agency (NEA) and Public Utilities Board (PUB), the public water association. NEA controls air and water pollution, handles waste, advances energy efficiency and environmental cleanliness, etc., whereas PUB deals with all matters concerning water in Singapore. To alert inhabitants for extending water-use efficiency, the public water utility sends water efficiency messages to individuals overall. The Singapore Power moreover gives flexible application that licenses inhabitants to see their noteworthy bills and portion status, gain better appreciation of the utility, and uses and submits meter readings. This leads buyers to audit their domestic use to manage their water consumption. In 2015, astute waste compartments were introduced as a piece of savvy squander organization program. The distinguishing screens joined on canister covers accumulate information on substance and region, and this is advised to a rubbish bunch through a focal server. These associate the waste combination gathering to smooth out their course organizing and, simultaneously, consistently keep the public spaces clean. The tainting levels are checked by NEA and are available for public on its position site. For model, occupants can, without very remarkable stretch, access information of 24-hour PSI regard, the planned air quality uncovering list on the web.

11.9.7 Energy

A brilliant technology-based Singapore expects to be energy proficient. With brilliant sensors inserted inside, domestic devices, for example, lights, can be put out automatically when nobody is at home. Excellent placing of lightings in buildings that can distinguish movement and change consequently is accessible.

For more effective energy use, Singapore's Wise Energy System (IES) endeavors to further develop network tasks and work with dynamic investment among shoppers. Shrewd meters that are outfitted with correspondence abilities assume a particularly fundamental part in permitting the framework to be two-way channel. They give the two buyers and the lattice administrator with data on how much power they use.

In 2012, as a piece of IES project under stage 2, savvy meters were introduced at about 1,900 families in Punggol. Some chose private families were given in-home showcase (IHD) units, which is a convenient gadget that gives families constant data about their power use. The pilot project was considered to make positive effects and, presently through brilliant meters, is set at home. Purchasers can screen their energy utilization, select their power retailer, just as picking their own energy bundle that best suits their necessities.

11.9.8 Citizen Interaction and Communication

As an associated country with incredibly noteworthy remote broadband association rate and savvy telephone entrance rate, the city has underscored the significance of availability among the Government, enterprises, and its individuals. The vast majorities of the Singapore Government's fundamental sites have grown more resident focal substance, and quite a bit of data are moved to residents through web-based media and versatile applications. Of course, Singapore is as of now an innovator in terms of its e-government stage. It has begun its excursion as ahead of schedule as the 1980s with the objective of transforming the Government into a world-class user of data innovation. In the 1990s, information innovation and telecommunications began to join, changing the idea of service conveyance. This has prompted the dispatch of the following e-Government designs as shown in the following paragraph.

Singapore's e-Government is controlled by the Ministry, and the IDA acts as the Chief Information Officer for the Government. iDA gives specialized guidance and recommendations, master planning, and project management administrations to MOF and other Governmental offices for the executives and implementation of e-Government services (programs).

E-Government programs are isolated into three classifications: programs for citizens, programs for organizations, and programs for government. Instances of smart city implementation using AI technology with major specifications are shown in Table 11.1.

TABLE 11.1
AI Technologies Used for Smart Cities

City Operations	Specification	Technology Used
City revenue management	#GIS-based property representation system #Secured, zone, and category-wise tax collection #Default tax #Analysis of multiple properties, and properties without name and number #Comparison of current and arrear tax	Image object detection Pattern recognition Trend forecasting Natural language processing
Mobility	#High-resolution capture of live traffic #Detect vehicles #Provide real-time data of mobility #Identify speed of vehicle and density of traffic	Camera-feed image object detection Pattern recognition Trend forecasting
Traffic enforcement system	#Number plate detection and violation monitoring #Video analytics of rule breaking #Speed violation detection	Camera-feed image object detection Image recognition
Smart parking	#Dynamic regulated parking slots during peak hours #Algorithm-based historical data #Monitors occupancy levels to adjust pricing in real time	Trend forecasting Traffic data image analysis
Smart street infrastructure	#Energy optimization of streetlights #Automatic on-and-off technology #Real-time calendar-based schedule #Monitors sectors-wise power consumption #Secures mesh network and automatic functionality in the case of lack of network connectivity	Sensor-based data forecasting Pattern recognition
Waste management	#Segregation of waste #Garbage collection #Efficient garbage management #Efficient resolution of complaints #Waste not collected alert	Image object detection Pattern recognition Trend forecasting Sensor-based data forecasting
Environment	#Forecasting #Effects on citizens #Water body health and contamination assessment #Carbon analysis #Impact of landfills #Estimating contribution of pollution #GIS assessment and warning indicator	Image object detection Pattern recognition Trend forecasting Sensor-based data forecasting
Safety and security	#Visualizes crime trends #Reports location #Predicts crime pattern	Pattern recognition

(Continued)

TABLE 11.1 *(Continued)*
AI Technologies Used for Smart Cities

City Operations	Specification	Technology Used
Urban planning	#Land utility #Land management system with ownership details #Assists to collect tax #Maintains data of property #Updates records regularly #Real-time information to manage utilities	GIS-based image object detection Pattern recognition Trend forecasting Natural language processing
Disaster and crowd management	#Visualizing location through geospatial solutions #Target disaster planning and effective communication #Impact analysis #Improve data distribution to responders and relief agencies #Track people density #Suggest alternatives	Dynamic optimization GIS-based image object detection Pattern recognition Trend forecasting Natural language processing

11.10 CONCLUSION

This chapter identifies various emerging themes and about sustainability in smart city development. The smart city concept is non-static and evolves over time. Sustainability-oriented definitions frequently center around the mix of delicate capital, such as human and social capital, and hard capital, a city's actual framework, to convey a practical, livable, and productive city. Then again, non-sustainability-oriented definitions as a rule featured the significance of ICT usage to effectively join assets that will make the city more interconnected, smart, and livable. Comprehensive methodologies covered all elements of sustainability, the natural, the social, and the monetary one and introduced a decent perspective on what a brilliant city ought to be. Generally, natural and social arranged definitions center on how savvy urban areas incorporate innovation with administration to work on the personal satisfaction and lessen the ecological effect of urbanism. Although the presence of unequivocal smart city procedures is lopsided across the world, the use of talks around AI and mechanical technology in these techniques is additionally restricted to only a couple of urban communities. As the arrangement of these advances, all things considered, turns out to be increasing far and wide, these spearheading urban areas fill in as forerunners of the developing and future inclinations. These urban communities articulate AI and mechanical technology: mechanization, navigation, instruction, brilliant foundation, and shrewd portability. Nonetheless, more exploration is required to deal with different social difficulties arising closeby the advancement of such innovations and practices in city.

REFERENCES

1. United Nations. Cities Will Continue to Grow Especially in Developing Countries. Available online: https://www.un.org/en/desa/around-25-billion-more-people-will-be-living-cities-2050-projects-new-un-report
2. Dameri, R.P. Smart City Implementation. In *Creating Economic and Public Value in Innovative Urban Systems*; Springer International Publishing: Cham, Switzerland, 2017; pp. 1–108.
3. Huang, S.L.; Chen, C.W. Theory of urban energetics and mechanisms of urban development. *Ecol. Model.* 2005, *189*, 49–71.
4. Caragliu, A.; Del Bo, C.; Nijkamp, P. Smart cities in Europe. *J. Urban Technol.* 2011, *18*, 65–82.
5. Silvestre, B.S.; Tirca, D.M. Innovations for sustainable development: Moving toward a sustainable future. *J. Clean. Prod.* 2019, *208*, 325–332.
6. D'Adamo, I.; Falcone, P.M.; Martin, M.; Rosa, P. A sustainable revolution: Let's go sustainable to get our globe cleaner. *Sustainability* 2020, *12*, 4387.
7. Zaheer, A. On big data, artificial intelligence and smart cities. *Cities* 2019, *89*, 80–91.
8. Banco de España. Escenariosmacroeconómicos de referenciapara la economíaespañolatras el Covid-19. *Bol. Econ. BancoEspan.* 2020, *2*, 1–35.
9. Bullough, A.; Renko, M. Entrepreneurial resilience during challenging times. *Bus. Horiz.* 2013, *56*, 343–350.
10. Elmqvist, T.; Andersson, E.; Frantzeskaki, N.; McPhearson, T.; Olsson, P.; Gaffney, O.; Takeuchi, K.; Folke, C. Sustainability and resilience for transformation in the urban century. *Nat. Sustain.* 2019, *2*, 267–273.
11. Roundy, P.T.; Brockman, B.K.; Bradshaw, M. The resilience of entrepreneurial ecosystems. *J. Bus. Ventur. Ins.* 2017, *8*, 99–104.
12. D'Adamo, I.; Rosa, P. How do you see infrastructure? Green energy to provide economic growth after COVID-19. *Sustainability* 2020, *12*, 4738.
13. Cocchia, A. Smart and Digital City: A Systematic Literature Review. In *Smart City. Progress in IS*; Dameri, R., Rosenthal-Sabroux, C., Eds.; Springer: Cham, Switzerland, 2014; pp. 13–43.
14. Osiyevskyy, O.; Shirokova, G.; Ritala, P. Exploration and exploitation in crisis environment: Implications for level and variability of firm performance. *J. Bus. Res.* 2020, *114*, 227–239.
15. Teece, D.; Pisano, G.; Shuen, A. Dynamic capabilities and strategic management. *Strateg. Manag. J.* 1997, *18*, 509–533.
16. Cruz, J.; López, P.; Martín de Castro, G. La influencia de lascapacidadesdinámicassobre los resultadosfinancieros de la empresa. *Cuad. Estud. Empres.* 2009, *19*, 105–128.
17. McElroy, M.W. Integrating complexity theory, knowledge management and organizational learning. *J. Knowl. Manag.* 2000, *4*, 195–203.
18. Corral de Zubielqui, G.; Jones, J.; Statsenko, L. Managing innovation networks for knowledge mobility and appropriability: A complexity perspective. *Entrep. Res. J.* 2016, *6*, 75–109.
19. Martín-Rojas, R.; García-Morales, V. J.; Garrido-Moreno, A.; Salmador-Sánchez, M. P. Social media use and the challenge of complexity: Evidence from the technology sector. J. Bus. Res. 2021, *129*, 621–640.
20. Su, K.; Li, J.; Fu, H. Smart City and the Applications. In *Proceedings of the International Conferences on Electronics, Communications and Control*, Nigbo, China, 9–11 September 2011; pp. 1028–1031.
21. Enerlis, Y. *Libro Blanco de Smart Cities*; Imprintia: Madrid, Spain, 2012; pp. 1–113.
22. Andalusia Smart Action Plan. 2020. Available online: https://www.juntadeandalucia.es/export/drupaljda/2016-08-30%20PAAS2020.pdf (accessed on 15 January 2020).

23. Elmahdy, H.N. *Medical Diagnosis Enhancements through Artificial Intelligence*; Ain Shams University Press: Cairo, Egypt, 2014; pp. 9–12.
24. Moursund, D. *Brief Introduction to Educational Implications of Artificial Intelligence*; University of Oregon Press: Eugene, OR, 2003; pp. 3–68.
25. Nordlander, T.E.; Nordlander, T.E. *AI Surveying: Artificial Intelligence in Business*; Montfort University Press: Leicester, 2001; pp. 33–54.
26. Reiterer, A.; Egly, U.; Björn Riedel, M.H. Application of Artificial Intelligence and Innovations in Engineering Geodesy. In *Proceedings of the Second International Workshop on Application of Artificial Intelligence and Innovations in Engineering Geodesy (AIEG 2010)*, Braunschweig, Germany, 7–9 June 2010; pp. 1–95.
27. Cristani, M. The complexity of reasoning about spatial congruence. *J. Artif. Intell. Res.* 1999, *11*, 361–390.
28. Mihael, H.C. *Artificial Intelligence*; Library Congress: Washington, DC, 1963; pp. 89–125.
29. LeCun, Y.; Bengio, Y.; Hinton, G. *Deep Learning*; Macmillan Publishers Limit: London, 2015; pp. 1–9.
30. Silver, D.; Aja, H. Mastering the game of Go with deep neural networks and tree search. *Nature* 2016, *529*, 484–489.
31. Giménez Figueroa, R.; Martín Rojas, R.; García Morales, V.J. Business Intelligence: An Innovative Technological Way to Influence Corporate Entrepreneurship. In *Entrepreneurship-Development Tendencies and Empirical Approach*; Mura, L., Ed.; InTechOpen: Zagreb, Croatia, 2018; pp. 113–132.
32. Knight, G.A. Cross-cultural reliability and validity of a scale to measure firm entrepreneurial orientation. *J. Bus. Ventur.* 1997, *12*, 213–225.
33. Zahra, S.A. Environment, corporate entrepreneurship, and financial performance: A taxonomic approach. *J. Bus. Ventur.* 1993, *8*, 319–340.
34. Daws, R. AI News. 2016. Available online: https://artificialintelligence-news.com/ (accessed on 13 July 2020).
35. Ramaswamy, S. How Companies are Already Using AI. 2017. Available online: https://hbr.org/2017/04/how-companies-are-already-using-ai (accessed on 13 July 2020).
36. Dodds, P.; Watts, D.; Sabel, C. Information exchange and the robustness of organizational networks. *Proc. Natl. Acad. Sci. USA* 2003, *100*, 12516–12521.
37. The University of Queensland. Blackboard Learn. 2015. Available online: https://learn.uq.edu.au/ (accessed on 19 May 2020).
38. Word Bank. Word Bank Open Data. 2018. Available online: https://data.worldbank.org/ (accessed on 15 January 2020).
39. World Bank. Urban Development. 2018. Available online: https://www.worldbank.org/en/topic/urbandevelopment (accessed on 25 January 2020).
40. World Health Organization. WHO Global Urban Ambient Air Pollution Database. 2016.
41. United Nations Department of Public Information. 2019. Available online: https://population.un.org/wpp/Publications/Files/WPP2019_PressRelease_ES.pdf (accessed on 26 January 2020).
42. United Nations Educational, Scientific and Cultural Organization. Urban Ex.
43. Martínez, J.M.S. Aglomeraciones y áreasurbanas en España, dimensión y tendencias: Brevesprecisiones. *LurraldeInvestig. Esp.* 2006, *29*, 115–142.
44. World Health Organization. Public Health, Environmental and Social Determinants of Health (PHE). 2016.
45. Ortega-Fernández, A., Martín-Rojas, R. and García-Morales, V.J. Artificial intelligence in the urban environment: Smart cities as models for developing innovation and sustainability. *Sustainability* 2020, *12*(19), 7860.

12 Security and Privacy Issues in Smart Cities

Sinimole K.R.[1] *and Soni Lanka Karri*[2]
[1]Information Systems & Business Analytics, Rajagiri Business School, Cochin, Kerela, India
[2]Faculty of Science, University Brunei Darussalam, Bandar Seri Begawan, Brunei Darussalam

12.1 INTRODUCTION

Smart cities indicate an intelligently connected settlement, responsible for ease of living, ease of administration and society enhancement and creation of happy communities. Consequently, many countries worldwide have incorporated the concept of a smart city into their community set-up. These countries have either constructed smart cities or are aggressively coining strategies to develop new cities which are either smart or converting their existing infrastructure to a smart city. However, *the ease* has been built into smart services through multiple junctures of processes and technology, with the Internet of Things (IoT) at the centre, which unfortunately also exposes these linkages to possible misuse by miscreants (Ismagilova et al., 2019; Singh et al., 2019). Consequently, success in executing and nourishing smart cities needs a rigorous framework with importance to security, privacy and trust rather than mere interrelated devices and algorithms.

Most importantly, the availability of smart technology has created a new challenge: "data" to harness and build upon. This "data" includes data already present in significant volumes, variety and velocity within the city databases and data generated through the smart links, sensors and other smart devices (Van Zoonen, 2016). The volume of data generated and issues related to data security, privacy and other cyber problems raise the question of when, where and how to address these criticalities (Awad et al., 2019; Elmaghraby & Losavio, 2014). It can also be noted that the threats, the unlawful access to data, arise at various stages of the creation of smart applications like modelling, development, deployment or at the execution stage.

12.2 SECURITY, PRIVACY AND PROTECTION ISSUES

According to Braun et al. (2018), security of the smart city, rather than an absolute concept, is a dynamic and genuine effort to avoid harm to the smart city both directly and indirectly and also through digital and physical connections (Braun et al., 2018). Despite the contributions made by smart applications and smart cities, their dependency on IoT technologies has made them vulnerable to hacking; hence, the privacy of the citizens is at stake (Bartoli et al., 2011; Eckhoff & Wagner, 2017). It's easy for hackers to breach and get personal information like habits, identity, locations

 DOI: 10.1201/9781003225317-12

and social behaviour. Applications related to the smart city, such as mobility, grids, healthcare, cards and smart home, are highly vulnerable to attacks, and personal data is highly likely to get leaked.

Smart mobility: Smart mobility applications, such as management of traffic flow and related activities, map reading, sensing the condition of the road, are designed to create safe and efficient road transportation (Ning et al., 2017). Nevertheless, the IoT and Internet of Vehicles (IoV) are at the centre of these mobility applications and hence are susceptible to known attacks and exposed to various innovative hacking methods. Consequently, the security aspects require more comprehensive and out-of-the-box thinking. For example, profuse traffic information may be used to presume users' locations and mobility patterns; hence, the users are unprotected from criminal activities. Any fleet data where each trip record is stored, if not well protected, is highly dangerous as attackers can view the location and tracks.

Smart home: The IoT is also at the heart of smart homes. Hence activities of the inhabitants and personal information like birth data, educational qualifications and credit cards details are captured and transmitted outside the house through cyber-attacks, implantation of spy devices or the complete modification of data. Moreover, smart home appliance makers and net distributors have subtle information, such as security codes, inhabitants' behaviours and daily routines. This information can be used to launch targeted attacks to rope the victim into shady deals. According to Alomair and Poovendran (2014), security systems based on transport layer encryption methods are insufficient to stop attackers from eavesdropping. The risk associated with smart homes is related to identity theft, spying and monitoring, location tracking and data manipulation.

Smart grids: Smart metering infrastructure is a complex constituent of smart grids for periodically recording optimised real-time power consumption. Like other smart home appliances, device manufacturers, service providers, generation plants and net providers have vital and sensitive information about consumers and utilities. Smart grids also monitor power flows, and because of this real-time monitoring, sometimes the privacy (living habits, working hours etc.) of residents get exposed to hackers or untrusted system operators (Finster & Baumgart, 2014). Based on the target, the attack on smart grids can be categorised into three primary categories: against the instrument, the manufacturer/creator/developer and the data transfer.

Smart healthcare: It's a fact rather than a prediction (Koop et al., 2008) that the delivery of healthcare services will soon be utterly home-centric from its present hospital-centric nature. The unexpected sudden outbreak of the COVID-19 pandemic has accelerated the idea of home-centric patient care. This home-centric patient care is possible only because of the developments in IoT and its applications in smart healthcare. Nevertheless, health being a personal and sensitive issue, the widespread use of smart applications in healthcare has challenged data privacy, trustworthiness and security and protection methods, and it's a real obstacle in the whole process (Gong et al., 2015; Yue et al., 2016).

Smart cards: Smart cards are a part of smart cities and are extensively used for the login process of various services, admission processes and transactions. However, smart cards collect sensitive and personal information (such as identity, behaviour patterns about cardholders and their daily routine) during these processes too. In addition, this vital information could be misused by distributors for pursuing, reporting and advertising (Eckhoff & Wagner, 2017).

In summation, smart applications have many advantages. However, they are imperilled to many vulnerabilities that concern the privacy of record owners too. Vulnerability, in a smart city, can be categorised as (1) detection of a particular person's record in the data, (2) identification of any of the record owners by scanning the data, (3) change of belief about sensitive attributes of a record owner and (4) inference of sensitive information about a record owner. The most severe existing attacks that intrude on privacy by exploiting vulnerabilities mentioned above are background knowledge attacks, linkability attacks, collusion attacks, Sybil attacks, eavesdropping attacks, man-in-the-middle attacks (MITM), neighbourhood attacks, skewness attacks, proximity attacks, freeform attacks, minimality attacks, spam attacks and spoofing attacks (Finster & Baumgart, 2014; Kitchin, 2016).

12.3 THE MOST SEVERE ATTACKS ON PRIVACY

Background knowledge attacks: An attack on the ***k-Anonymity privacy model***[1] is one of the most widely used methods to breach an entity's personal information. The attacker can identify personal information from the anonymised table if the background information (such as people's email and age) about the target (specific people in smart cities) is available, thus resulting in attribute linkage or attribute disclosure (Machanavajjhala et al., 2007). The information acquired through this process is referred to as "post-belief". A successful attack is confirmed if this post belief contains more subtle information than the initial belief (Gil-Garcia, 2012). The difficulty in modelling the prior assumptions makes the background attacks complex and challenging to measure adversaries.

Linkability attack: Linkability attacks which result in record linkage or identity disclosure occur when an attacker combines released data with background knowledge to identify record owners (Abdallah & Shen, 2016; Sweeney, 2000). It makes use of sensitive information shared by the users while creating various linkages in the communication network corresponding to different nodes and stakeholders (Yu et al., 2016). Proficiency in machine learning (ML) algorithms and their uses in data synthesis make it easy for attackers to take control of user data and launch linkability attacks (Abdallah & Shen, 2016).

Collusion attack: Amongst the attacks, collusion attack has become one of the fast-growing threats to data privacy. Collusion attacks generally occur in multiple-party set-ups. It is the execution of operations such as averaging, replacement or linear combination of media copies to create new copies and to gain more sensitive knowledge about citizens (Rezvani et al., 2015). It is not compulsory to have the background or prior knowledge of the inhabitants (Tao et al., 2016). However, the previous knowledge about the inhabitants acts as an added lead to carry out the attack easily (Koufogiannis & Pappas, 2016). If the adversary doesn't have any prior information, then the attacker combines different k-anonymous versions of the data to gain information about the users. Collusion attacks also result in identity disclosure. The easiest way to prevent a collusion attack is to keep the privacy level uniform for all the nodes.

Sybil attack: This attack occurs when the system's reputation is subverted by forging identities in peer-to-peer networks by creating a large number of pseudo-names to gain maximum influence (Quercia & Hailes, 2010). The collected information is used to deceive the inhabitants and the authorities. As the cost of creating fake identities is a matter of concern, whether an attack is successful (success rate of an attack) is determined only after the cost-benefit analysis. Generally, the adversaries never lose a chance to reduce the cost of fake identities; for example, fake identities can be created easily in a disaster where citizens rush together to help the victims.

Eavesdropping attack: An eavesdropping attack is also known as a sniffing or snooping attack. Attackers get access to the communication channel and usually eavesdrop (listen) on the data transmission process as the data is communicated through the interrelated network by connected devices (like smart healthcare devices, smart grids, computers, smart watches, smartphones and so on). The attackers use unsecured network communications to passively listen and gain access to private information, such as birth information, health data, node identification numbers, travel updates and so on, as it is being sent or received by its user. What makes it more complicated is that it is not easy to identify this attack, as the existence of a breaching or listening device may not harm the performance of the connected devices and the network as a whole. The attacker can use the eavesdropped data to extort the nodes to negotiate, interrupt the communications between the nodes or reduce the system's efficiency.

A man in the middle attack (MITM): This is a type of eavesdropping attack, where attackers interrupt an existing conversation or data transfer by faking the identity of a network node or device and pretending to be legitimate participants. This enables an attacker to capture or stop information and data from either party while sending malevolent links or other information to authentic participants in a way that might not be detected until it is too late. This type of attack can be prevented by using cryptographic protocols to ensure secure communication. Many public networks do not use encryption when exchanging data, thus making it easy for attackers to access the system (Awad et al., 2019). Policies need to be devised to allow for suitable data communication standards for such networks while keeping the user as well as commercial interests in mind.

Neighbourhood attack: According to Zhou and Pei (2008), when an attacker who knows about the neighbourhood of a person and the relationship among neighbours uses this information to collect the identity of an inhabitant and uses it to harm the person through an online social network, the neighbourhood attack occurs.

Skewness attack: When an attacker guesses the sensitive information of the target using the similarity or the relationship among the values in an anonymous table, the skewness attack occurs (Li & Li, 2007). This attack is called a skewness attack because it is induced by the attacker using skewed sensitive values. According to Li and Li (2007), this type of attack results in attribute linkage.

Proximity attack: This attack is similar to the skewness attack. Here, the attacker guesses the sensitive information with confidence when some values within a range are frequently present in the microdata (Li et al., 2008). This attack, too, results in attribute linkage.

Freeform attack: Freeform attack is another attack which guesses a sensitive attribute with confidence (Wang et al., 2009). The attacker uses any attribute of a record owner, not necessarily quasi-identifier, in microdata to guess the sensitive value. The freeform attack also results in attribute linkage.

Minimality attack: This kind of attack uses knowledge about the anonymisation algorithms to infer sensitive information (Wong et al., 2007).

Spam attack: In spam attacks, the attacker sends emails to inhabitants whose email addresses are obtained in advance and further spies on the storage space. A proposed system named "LENS" (LEveraging anti-social Networking against Spam), based on the anti-social networking paradigm, can address spam attacks and prevent their transmission using gatekeepers (Hameed et al., 2011).

Spoofing attack: In a spoofing attack, an adversary becomes part of the network and pretends to be a legitimate device on the network, hence getting the opportunity to send irregular or abnormal data to upset the regular operation of a smart city system. As the IoT devices in a smart city have varying levels of built-in security, spoofing can seriously harm the system. The use of cryptography, hybrid encryption as well as the use of blockchain (Xiao et al., 2016) can be used to prevent spoofing attacks by validating data exchange and authenticating devices.

In addition to the previously mentioned attacks which are the most severe ones, other significant issues are smart devices monitoring the private life (work timings, living habits etc.) of citizens (Ning et al., 2017), manufacturers and distributors of smart instruments getting a hold on data (Tan & Wang, 2010) and the profound use of information collected by the smart traffic system monitors to infer location and mobility patterns of a user (Petrolo et al., 2017). In addition to this, to aggravate the situation, rapidly developing smart applications like the ones mentioned in the following also create a threat to privacy.

12.4 THREATS DUE TO FAST-DEVELOPING TECHNOLOGY

Botnet: Recent technological developments have created a wide range of prospects for both consumers and companies. However, this comes at a cost because it gives fraudsters the chance to get illegal access. A botnet is a smart collection of web-connected devices that are used to conduct cyberattacks. Attackers or cyber-criminals use botnets as a forced multiplier to disrupt or break into their targets' systems. Sending spam, stealing data, utilising private information or launching severe distributed denial of service (DDoS) attacks are all common botnet attack tactics. The proliferation of devices is one of the most frequent obstacles to preventing these attacks. Monitoring, sensing and stopping botnet attacks are getting harder as more and more devices become widely available. As the concept is new, an efficient defence mechanism is not currently available and many more studies are needed to develop an efficient novel approach (Yu et al., 2015).

Driverless cars: Another threat, which is dangerous, is the rapidly growing application of autonomous vehicles (AVs). AVs need extreme care and precaution because once attackers get a hold of the control of AVs, the life of citizens and data security is under threat (Chiang & Zhang, 2016). It can happen in two ways: One is by attackers conducting remote attacks (examples are applying brakes, shutting down the engine and controlling the steering). The second is by misusing the massive data collected by the computer system.

Virtual reality: Virtual reality (VR) technology is a part of smart cities nowadays. Many organisations in the healthcare sector, engineering sector, planning sector etc. have incorporated VR technology in their day-to-day activities. But the sensitive data shared by these departments with external devices or persons, the unencrypted data exchange between VR devices and the information stockpiled by sensors all contribute to privacy leakage (Cui et al., 2018). Lamentably, the designers have not taken adequate care of privacy issues as they were in a hurry to rush their products to the market.

Artificial intelligence (AI): AI has a central role in various smart applications. Nevertheless, the use of AI technologies creates security issues too. The stakeholders (like service providers and manufacturers) who have access to these systems and attackers can use various AI techniques to analyse personal information and extract sensitive information for their selfish reasons (Acquisti et al., 2015; Li et al., 2011). Attackers are good at analysing ML algorithms, and thus, they can collect sensitive information without much effort. They can weaken the algorithms' reliability too, allowing other hackers to snoop into the system easily.

In summation, we have learned about the vulnerabilities in smart cities, the most severe attacks caused by exploiting these vulnerabilities, significant problems in different application scenarios and the new threats due to the developments in technology related to smart applications. Considering the multifaceted nature of smart cities and the data threats mentioned earlier, the prerequisites to the protection mechanisms are knowledge about smart cities' characteristics and security requirements. A smart city is characterised by a network of infrastructure based on highly heterogeneous, resource-constrained technologies. As far as security requirements are concerned, authentication and confidentiality, availability, integrity, intrusion detection and prediction are the important ones (Midi et al., 2017). The following section describes a smart city's characteristics and security requirements.

12.5 CHARACTERISTICS AND SECURITY REQUIREMENTS

12.5.1 Characteristics

Heterogeneity: This means that the systems in smart cities are independent, distributed with various subsystems and resource-constrained nodes and stockpiled or availed by different inhabitants. Heterogeneity also denotes the complex communication protocols and technologies, a wide variety of network connectivity options, mobility means, various interrelated devices, platforms etc. (Jing et al., 2014) which are sometimes unknown to a service provider.

Resource constraints: IoT devices (such as sensors and smart devices etc.) are resource constrained. These constraints include inadequate CPU, memory, battery capacity and processing capabilities. Additionally, their network interfaces are constrained due to low-power radio standards, among other factors. Because of these limitations, the security features are not fool-proof. They are small, inexpensive and low-powered embedded devices with limited computational capabilities. A typical sensor has a minimal memory, either 8- or 16-bit, which is not enough for a smart application to work.

Mobility: Understanding the mobility of smart city devices is essential to improve quality of service in smart applications. Mobility of smart devices refers to an intelligent transportation network through which the instantaneous monitoring of the trajectory and flexible communication and transportation of the devices and inhabitants take place. That means mobility is not just the transportation of goods and vehicular movements within the smart city.

Connectivity: This characteristic empowers smart devices, which are rapidly becoming an indispensable part of our life, to connect. Smart applications are growing in number, and hence the connectivity between them (in terms of the volume and transition of data and vehicular movements) is rapidly increasing. Therefore, it's essential to have an efficiently and effectively connected network.

12.5.2 Security Requirements

Authentication: Authentication is very important for ensuring security and is an elementary necessity of a smart system. A smart application may have many operators, and the different operators may have different access privileges. Proper authentication and access control are required at various levels. All smart applications must prove identities and guarantee that only authorised users can enter the heterogeneous system (He et al., 2016). Specifically, smart devices need to validate the network, the resource-constrained nodes and the communications from administrators. Regrettably, most existing authentication mechanisms are inadequate and inefficient in safeguarding the system against security and privacy threats. It is crucial to address these shortcomings promptly to ensure better protection.

Confidentiality: The term confidentiality refers to the protection of data from submissive attacks and also from getting exposed to a wrong choice. In smart applications, as mentioned earlier, invaders have the capability to eavesdrop on communication or access devices. Therefore, encryption-based technologies are widely applied to protect the confidentiality of information transmission between nodes (Angelakis et al., 2017). Notably, transparency and reliability are two factors that make the design of identification and authentication methods difficult (Zhang et al., 2015).

Availability and integrity: Smart systems or applications should be available and effectively functioning even under attack; that is, the system should be resilient enough to tolerate various faults and failures caused by the attackers. Moreover, because these devices are vulnerable to attacks, they should be able to sense the anomalies and have the capability to prevent further impairment to the system. Smart devices also should have the mechanism to automatically adapt to the environment to cope with increasingly intelligent attacks. In an interconnected network, data must be exchanged across the devices frequently; therefore, there is a high possibility of data getting altered during the transmission process if they are not well protected. Some methods can effectively control the data flow between the interconnected devices. However, these methods fail to assure the data integrity at nodes, especially at the remote ends, as most IoT devices' computational power is very low.

Detection and prediction: A secure system is one which has the ability to monitor its operational environment and has the ability to sense abnormalities promptly. The conventional systems for identifying threats are generally conditioned to notice embezzlers and to sense anomalies (Abduvaliyev et al., 2013). However, in the assorted, complex and resource-constrained smart environments, the simple conventional detection methods are not recommended and are impractical (Midi et al., 2017) and hence need lightweight intrusion detection methods.

Many devices, especially web-based devices, are not successfully detecting and preventing attacks in advance (Xynos et al., 2013). Most of the time, the attacks come to the light only after they cause severe damage to the system; sometimes irreparable and current protection methods are incapable of safeguarding the system (Wu et al., 2016). Thus, it becomes essential to have intelligent systems to predict attacks on smart applications automatically.

12.6 CURRENT SECURITY AND PRIVACY PROTECTION TECHNOLOGIES

As discussed throughout the chapter, privacy protection, attacks or threats and sensitive data leakage (deliberate or accidental) are the causes of privacy breaches and have become the main problems in our data-driven smart world. Extant research (Cui et al., 2018; Zhang et al., 2017) on data privacy identifies four data sources from which the hacking can take place. They are observable, repurposed, published and leaked, containing many users' sensitive information (Cui et al., 2018; Zhang et al., 2017). Therefore, to avoid misuse and hacking, adequate and effective countermeasures, such as encryption methods, clustering-based methods and some novel techniques, such as differential privacy (Catarinucci et al., 2015; Dwork et al., 2006; Magno et al., 2015; Mudumbe & Abu-Mahfouz, 2015) must be applied. Differential privacy is a rigorous method which has attracted increasing attention and applications recently.

The techniques currently used to prevent security and privacy threats in smart cities are cryptography, biometrics, blockchain, ML and data mining, game theory, ontology, differential privacy and some nontechnological strategies.

12.6.1 Cryptographic Methods

Cryptographic methods are widely used for the protection and prevention of security and privacy threats in the connected world. Even though they are considered the backbone of security and privacy protection algorithms, traditional encryption methods, which are computationally complex, are not entirely appropriate for resource-constrained smart devices (Jing et al., 2014). Therefore, lightweight encryption methods that perform well in a constrained environment are considered a prerequisite for the application of cryptographic algorithms in smart devices.

Despite the richness of algorithms and their competence, lightweight encryption requires copious numbers of research and solutions because of the difficulty in attaining the required level of security. One such method which attracts increasing attention is homomorphic encryption (HE), a method of executing calculations on encrypted data. Examples are as follows: HE has been used in medical computations (Sun et al., 2017), smart healthcare monitoring systems (Talpur et al., 2015), smart grids (Sehgal et al., 2012) and cloud computing (Jabbar & Najim, 2016). However, HE methods are computationally expensive and therefore are restricted in their use.

Another cryptographic method which is in use in smart city applications is "zero-knowledge proof". Dousti and Jalili (2016) used "zero-knowledge proofs" to develop an authentication protocol for smart cards. There are many more applications of lightweight cryptographic methods to protect privacy, such as developing an authentication mechanism for protecting end users' communications from DDoS attacks (Mahmood et al., 2016), public key encryption scheme (Li et al., 2017) and so on.

12.6.2 Biometrics

Most smart applications use biological features such as voices, irises, fingerprints, faces, brainwaves, handwritten signatures and so on to enter and authenticate the smart applications, which is a simple mechanism to protect the personal data of inhabitants from attackers. Out of the biometric methods mentioned before, brainwave-based authentication (Zhou et al., 2017) is an efficient and accurate authentication method. Protection based on biometric methods is more robust than traditional password-based methods. However, the data generated by the biological methods are also sensitive, and if it is not used correctly, it can also lead to serious privacy issues (Natgunanathan et al., 2016; Yu, 2016). As a solution, Wang et al. (2017a,b) developed a privacy-preserving biometric method, making it difficult for attackers to deduce identities. Amin et al. (2017) developed a novel, cost-effective and efficient key negotiation and verification protocol to protect users' confidential data in storage devices.

12.6.3 Blockchain

Blockchain is a decentralised, realisable and significant technique that controls the data flow between smart devices to protect privacy (Christidis & Devetsikiotis, 2016). There are many applications of blockchain technology in smart city-related applications. For example, Yue et al. (2016) used blockchain technology to develop a novel healthcare system which protects personal healthcare data from attackers and hackers. Similarly, Dorri et al. (2017) developed a system based on blockchain technology to promise security to the smart automotive industry and to protect smart vehicle users from privacy attacks. One other example is the security framework developed by Biswas and Muthukkumarasamy (2016). It is a reliable and efficient framework to ensure the security of smart devices in a smart city network. Even though the applications of blockchain technology to protect privacy and security are promising, the potential of blockchain technology has still not been utilised fully. Therefore, smart application developers should take thoughtful measures to ensure the maximum utilisation of this powerful tool.

12.6.4 Machine Learning and Data Mining

ML techniques have been commonly used to improve the efficiency of algorithms used for threat detection and protection of networks. Some applications used an ML-based method to secure data sensing, a novel feature extraction and selection model, to detect attacks in Wi-Fi networks (Luo et al., 2016). Aminanto et al. (2017) predicted the privacy preferences of mobile users when using smart applications using ML techniques.

Wijesekera et al. (2017) tried to prevent personal data from being sent to any distrusted entities in the cloud. Rodriguez et al. (2017) used a support vector machine algorithm–based multi-sensor authentication system for smartphone users. Lee and Lee (2015) developed a novel permission mechanism for mobile platforms. A game-theoretic model can also detect and prevent interruptions (Olejnik et al., 2017; Shamshirband et al., 2014).

Service providers and government agencies use data mining technologies to improve the quality of services (Tsai et al., 2014). However, data mining also creates serious privacy concerns as it involves collecting and sharing sensitive knowledge that adversaries can extract, such as social styles, behaviour patterns and even religious beliefs. As data mining techniques can both be friend and foe, they should be used sensibly (Xing et al., 2017).

12.6.5 Game Theory

This is a powerful management science tool that has been effectively used to analyse strategies. Recently this has been widely applied to provide cybersecurity, especially in smart applications. The characteristics like a robust mathematical base, consistency in the results, and appropriate and distributed solutions make game theory-based models more effective than the traditional methods (Cuong et al., 2019). Some examples in which game theory is successfully used are game theory techniques based on the Bayesian approach to develop a k-anonymity method to hide the location details of the users from adversaries (Liu et al., 2013), a privacy trade-off model based on differential privacy and game theory to provide security to smart grids (Yassine et al., 2015), a game theory–based new strategy for exploring security violations in cloud storage (Abass et al., 2017), a game theoretic model to study the attack and defence problem in honeypot-enabled networks (La et al., 2016), a game theory model based on honeypot strategy to address the issues related to security threat in smart metering infrastructure (Wang et al., 2017a,b), "zero-sum game theory model" to protect the wireless networks from spoofed attackers, a protection model which is a combination of traditional methods, game theory, and the concept of "k-anonymity" (Liu et al., 2013) and differential privacy (Kearns et al., 2014) and so on. Most importantly, as game theory is a multicriteria decision-making model to study strategic behaviour, it can effectively balance protection intensity and data utility. Even though game theory approaches have been widely used in IoT devices, applications specific to smart cities are limited in the literature. We certainly hope that the designers of smart applications will realise their strong base in mathematics and adopt this tool to manage the threats in smart cities successfully.

12.6.6 Ontology

Ontology as a concept is a branch of philosophy. However, recently it has been widely used to formulate knowledge for tackling heterogeneous privacy-related issues. Models based on this concept work well when the data is not structured. It can be used to identify the formally represented data, knowledge or information that group together and the reason of their grouping, thus isolating the inconsistencies. This has helped the inventors to develop ontology-based models to resolve security and privacy problems, such as cyber-attack detection and security risk management (Mozzaquatro et al., 2015; Razzaq et al., 2014). Some examples are as follows: Tao et al. (2016) proposed an ontology-based security management model for smart devices to interact more effectively with other resource-constrained nodes. The ontology-based framework developed by Mohsin et al. (2017), which helps in the automatic capturing of consistencies in the process of interactions, is an excellent addition to security analysis. Quality of private information (QoPI) is an ontology-based model that Kim et al. (2017) developed for managing and characterising mobile users' data from possible pattern-finding attacks. Ontology-based models have great potential in solving trust-related issues and can do miracles in the "trustworthiness of smart applications". The term "trust ontology" is developed by Lee et al. (2017) and used to measure the trustworthiness among content providers and consumers according to users' preferences, purposes and perspectives. Despite its benefits, there are certain drawbacks: for example most ontology-based models focus on specific areas and lack an integrated model. To a certain extent, specifically at the network layer of IoT, the semantic ontology-based "situation reasoning model" developed by Xu et al. (2017) addresses this issue.

12.6.7 Differential Privacy

Many heuristics are employed to protect users' privacy when using smart devices. The most frequently employed method is anonymisation, which can fail and has been repeatedly demonstrated to fail. For example, it is possible to identify specific health records by linking attributes like gender, date of birth and zip code from a supposedly anonymised healthcare database to public voter information. These linkage attacks emphasize the need for a robust privacy concept that can withstand attacks utilizing auxiliary knowledge, especially when the data curator cannot anticipate its availability. Differential privacy is a more modern and effective strategy that solves several problems that were a part of the older methods. Its popularity has substantially increased due to its strict guarantees for upholding the privacy and security of the data. A prerequisite for applying differential privacy is the division of attributes into sensitive, non-sensitive and ambiguous groups. Sensitive attributes are not saved in the cloud as they contain personal information like names and other distinctive identifiers. Non-sensitive assets are stored on the cloud because they don't include personal data.

Differential privacy is a powerful and robust mathematical tool widely used in ML algorithms to solve privacy issues. Models based on differential privacy work well in preventing differential attacks, linkage attacks and reconstruction attacks. One of the well-known examples of differential privacy is the Laplace mechanism, in which Laplace-distributed noise is added to personal information to prevent privacy attacks.

12.6.8 Nontechnological Strategies

Nontechnological strategies are countermeasures based on policy formulations, awareness programmes, training and educating the stakeholders. Policies mostly contain guidelines and standard operating procedures, revision requests, verification, privacy notices, anti-retaliation policies and so forth. Government policies and regulations have a significant role in preventing privacy violations and controlling attackers. For example, training empowers smart application developers to create powerful models with sophisticated privacy controls. Furthermore, education empowers smart city inhabitants with basic data privacy and protection techniques (Hurst et al., 2017); hence, they will not ignore privacy concerns for the benefits and convenience of smart services (Aleisa & Renaud, 2017).

In summation, the application of methodological solutions alone is insufficient for protecting privacy; sound governance is critical in creating a reliable smart system. Prevailing technology limitations can be moderated by robust policies, regulations, administration and education (Kitchin, 2016; Meijer & Bolívar, 2016). The first and foremost one is the regulations enforced by the government. It should be able to protect the data and the security mechanisms in smart cities (Batty et al., 2012; Walravens, 2012). The second important factor is the training given to all the stakeholders, including, but not limited to (1) the dos and don'ts for the end users, (2) improving the skills of developers and distributors, (3) efficient administration and so on (Hurst et al., 2017). To cope with the rapidly changing technology, developers and designers should be trained to make a stable and resilient system.

Moreover, it is equally important to update the system (e.g., update the firewalls) as per the technological developments to fix vulnerabilities. Another important aspect is to enhance the overall safety, and quality standards and the device manufacturers are responsible for it. The next important stakeholder is the citizens. There should be programmes conducted to educate the citizens about the functioning of smart applications and their protection mechanisms (Perera et al., 2015). Upon analysing the current trends, it can be seen that the effectiveness of such programmes is very low. According to the researchers, many users ignore most of the warnings the protection mechanisms give, even though they know the seriousness of the privacy leakage, and let the attack happen (Aleisa & Renaud, 2017). This is an issue which needs immediate attention.

12.7 PERSPECTIVE: THE TRADE-OFF BETWEEN PRIVACY AND DATA UTILITY

The strength of smart cities lies in the fact that they provide improved quality of life to the citizens, which is operationally sustainable and less costly. Moreover, the real-time information collected by the devices helps the system function efficiently and provide better services to the citizens, thus contributing to better decision-making. However, on the other hand, as discussed in the chapter, the data collected generates severe security and privacy issues too. As we have seen in the previous sections, traditional security methods might not be sufficient or applicable due to their resource-constrained nodes, consequently exacerbating the privacy and security concerns among the inhabitants. Thus "trust" is the primary concern of smart citizens. They are always sceptical about managing the data, especially how the companies or the organisations providing the services will collect their data. Who is responsible for creating policy guidelines, standard operations procedures or benchmarks so that the inhabitants can use the smart applications freely? Is it government or the service providers? Or are the existing data policies or frameworks mature enough to regulate data management? Will transparency in managing the data help to improve trust? If so, what could be the level of transparency? Unfortunately, the real problem smart cities face is the **lack of data policies and legislation to protect data privacy in a connected world**.

To a certain extent, transparency will help increase the level of trust and enable companies to develop better data privacy mechanisms. Personal data transparency includes the kind of data stored and its accessibility and protection. Transparent mechanisms increase the trust levels of the inhabitants by helping them know more about the acquisition of data, data aggregation and retention methods, preservation period, protection mechanisms and their efficacy and so on, hence increasing the trust levels. Sometimes users might not agree with the data collection process. In such scenarios, trust can be built by giving options to the users (choosing policies) and the freedom to say which part of the data collection process is not acceptable to them.

Another vital aspect readers should focus on is misunderstanding protection levels and their durations. In reality, the level of protection does not remain constant all over the life cycle of the mechanism. It doesn't have an unchanging protective standard; it is personalised and changes as per customer needs. As most services are personalised, assuming that everyone needs the same level of protection is unfair.

Similarly, it is dangerous to believe that all the attackers have the same level of expertise and background knowledge. Such assumptions will definitely impact the quality of services provided and the usefulness of the data. Therefore, in the world of smart applications, "**personalisation**" is the key to enhanced trust and privacy. Personalisation improves the data utility too. The developers of smart applications can think of various proactive methods to enhance trust. They can embed the idea of protection mechanisms at the design stage instead of taking action after privacy attacks. In summation, there are many ways to build up trust among users. They can protect for a prolonged period or even the entire data life cycle.

Moreover, as discussed in the chapter, the current deployment scenario has a myriad of different technologies and software frameworks that do not often integrate well, again pointing to the minimum standards that the service providers must mandatorily keep. Besides, there are different standard bodies all over the world, for example, the Institute of Electrical and Electronic Engineers (IEEE), the European Telecommunications Standards Institute (ETSI) and the Internet Engineering Task Force (IETF). All these bodies have been contributing with different standards for data management, network communication and managing resource-constrained nodes. Unfortunately, many of these standards do not work well together. They cannot adequately address interoperability issues independently, nor are they compatible. Thus, smart cities present many opportunities for researchers and developers to mitigate weaknesses, produce compatible devices and develop new devices and services. Moreover, the data gathered also gives wonderful opportunities for developing new data science algorithms. For example, algorithms can be created for service delivery usage, computationally cheap encryption techniques, efficient data storage methods and networking technologies to make smart city deployment safer, trustworthy, more accessible and affordable.

12.8 CHALLENGES AND FUTURE DIRECTIONS

The challenges, security breaches and privacy protection mechanisms in smart cities today have been discussed throughout this chapter. We have also discussed many novel countermeasures recently proposed in various fields. However, current protection mechanisms are not enough to prevent advanced threats. Therefore, there is more space to accommodate effective and efficient protection tools to catch up with rapid technological growth. The readers are asked to explore the following challenges/promising opportunities for further advancements.

Security and privacy issues in fog-based systems: The fog-based system is an emerging technology which is nothing but a decentralised computing structure which allows the users to place the data or resources anywhere between the cloud and the smart applications to enhance the performance of the smart devices. Fog computing is a highly virtualised platform that offers networking, computing and storage services between end users and typical cloud data centres. The objective of fog-based systems is to provide analytical services at the edge of the network so that the distance through which the data transmission takes place reduces drastically, improving efficiency and enhancing privacy protection (Kolias et al., 2017). Some of the properties of fog computing, location awareness and low latency, end-device mobility, the ability to process many nodes, wireless connectivity, real-time applications and heterogeneity have quickened the analytical services. Fog computing increases system efficiency by tightly integrating with front-end smart end devices, which boosts the functionality of crucial cyber-physical systems. Fog computing organises and administers the local virtual cluster, whereas cloud computing strives to optimise resources from a global perspective.

However, structures raise new challenges and are vulnerable to attacks (Bastug et al., 2017) because it allows users to place data anywhere between the cloud and the device to improve its utility. Moreover, these systems are small and limited in capacity. The protections of smart devices in fog-based smart systems should receive much more attention (Kolias et al., 2017).

Lightweight security solutions: Even though there are plenty of methods for privacy protection, many of them are not compatible with smart applications because of their limited processing and computational powers. Traditional cryptography is created at the application layer. Therefore, it becomes essential to develop lightweight mechanisms for privacy protection considering the requirements like flexibility, dynamic and low cost. Recently, many studies have been put into developing lightweight cryptography for IoT. A type of encryption made for resource-constrained devices is lightweight cryptography. Radio frequency identification (RFID) tags, sensors, contactless smart cards and medical equipment are a few resource-constrained devices where lightweight cryptography can be used. Lightweight encryption technology employs less memory, less processing power and fewer computing resources to provide secure solutions to resource-constrained networks. For these reasons, it would be preferable to use lightweight security solutions.

Theoretical complement: The concept of smart applications, smart homes and smart cities is there worldwide. Every country has standards and protocols; accordingly, they have either developed their smart cities or projects are under development. That means the smart city concept is not uniformly defined; it varies from country to country. Consequently, the protection mechanisms and protocols which are currently available cannot be uniformly applied to all smart cities. Therefore, additional theoretical studies are necessary to reduce the barriers to securing smart cities.

NOTE

1 k-anonymity aims to prevent an attacker from linking released data with background knowledge and matching fewer than k record owners.

REFERENCES

Abass, A. A., N. Mandayam, and Z. Gajic, (2017). Evolutionary game theoretic analysis of advanced persistent threats against cloud storage, *IEEE Access* 99, 1–1. https://doi.org/10.1109/ACCESS.2017.2691326

Abdallah, A., and X. Shen, (2016). A lightweight lattice-based homomorphic privacy-preserving data aggregation scheme for Smart Grid, *IEEE Trans. Smart Grid*, 9, 396–405.

Abduvaliyev, A., A. S. K. Pathan, J. Zhou, R. Roman, and W. C. Wong, (2013). On the vital areas of intrusion detection systems in wireless sensor networks, *IEEE Commun. Surv. Tutor.*, 15(3), 1223–1237.

Acquisti, A., L. Brandimarte, and G. Loewenstein, (2015). Privacy and human behavior in the age of Information, *Science*, 347(6221), 509–514.

Aleisa, N., and K. Renaud, (2017). Yes, I know this IoT device might invade my privacy, but I love it anyway! A study of Saudi Arabian perceptions, in *Proceedings of the 2nd International Conference on Internet of Things, Big Data and Security, IoT BDS*, Porto, Portugal, April 24–26, 198–205.

Alomair B., and R. Poovendran, (2014). Efficient authentication for mobile and pervasive computing, *IEEE Trans. Mob. Comput.*, 13(3), 469–481.

Amin, R., R. S. Sherratt, D. Giri, S. H. Islam, and M. K. Khan, (2017). A software agent enabled biometric security algorithm for secure file access in consumer storage devices, *IEEE Trans. Consum. Electron.*, 63(1), 53–61.

Aminanto, M. E., R. Choi, H. C. Tanuwidjaja, P. D. Yoo, and K. Kim, (2017). Deep abstraction and weighted feature selection for Wi-Fi impersonation detection, *IEEE Trans. Inf. Forensics Secur.*, 13(3), 621–636.

Angelakis, V., E. Tragos, H. C. Pöhls, A. Kapovits, and A. Bassi, (2017). Designing, developing, and facilitating smart cities, in *Urban Design to IoT Solutions*. Springer.

Awad, A. I., S. Furnell, A. M. Hassan, and T. Tryfonas, (2019). Special issue on security of IoT-enabled infrastructures in smart cities, *Ad Hoc Netw.*, 92. https://doi.org/10.1016/j.adhoc.2019.02.007

Bartoli, A., J. Hernandez-Serrano, M. Soriano, M. Dohler, A. Kountouris, and D. Barthel, (2011). Security and privacy in your smart city. In: *Proceedings of Barcelona Smart Cities Congress 2011*, Barcelona, Spain.

Bastug, E., M. Bennis, M. Médard, and M. Debbah, (2017). Toward interconnected virtual reality: Opportunities, challenges, and enablers, *IEEE Commun. Mag.*, 55(6), 10–117.

Batty, M., K. W. Axhausen, F. Giannotti, A. Pozdnoukhov, A. Bazzani, M. Wachowicz, G. Ouzounis, and Y. Portugali, (2012). Smart cities of the future, *Eur. Phys. J. Spec. Top.*, 214(1), 481–518.

Biswas K., and V. Muthukkumarasamy, (2016). Securing smart cities using blockchain technology, *2016 IEEE 18th International Conference on High Performance Computing and Communications; IEEE 14th International Conference on Smart City; IEEE 2nd International Conference on Data Science and Systems (HPCC/SmartCity/DSS)*, Sydney, NSW, Australia, IEEE, pp. 1392–1393, https://doi.org/10.1109/HPCC-SmartCity-DSS.2016.0198

Braun, T., B. C. M. Fung, F. Iqbal, and B. Shah, (2018). Security and privacy challenges in smart cities, *Sustain. Cities Soc.*, 39, 499–507. https://doi.org/10.1016/j.scs.2018.02.039

Catarinucci, L., D. De Donno, L. Mainetti, L. Palano, L. Patrono, M. L. Stefanizzi, and L. Tarricone, (2015). An IoT-aware architecture for smart healthcare systems, *IEEE Internet Things J.*, 2(6), 515–526.

Chiang M., and T. Zhang, (2016). Fog and IoT: An overview of research opportunities, *IEEE Internet Things J.*, 3(6), 854–864.

Christidis, K., and M. Devetsikiotis, (2016). Blockchains and smart contracts for the Internet of Things, *IEEE Access*, 4, 2292–2303.

Cui, L., G. Xie, Y. Qu, L. Gao, and Y. Yangi, (2018). Security and privacy in smart cities: Challenges and opportunities, *IEEE Access*, 6, 46134–46145, https://doi.org/10.1109/ACCESS.2018.2853985

Dorri, A., M. Steger, S. S. Kanhere, and R. Jurdak, (2017). Blockchain: A distributed solution to automotive security and privacy, *IEEE Commun. Mag.*, 55(12), 119–125.

Dousti, M. S., and R. Jalili, (2016). An efficient statistical zero-knowledge authentication protocol for smart cards, *Int. J. Comput. Math.*, 93(3), 453–481.

Dwork, C., F. McSherry, K. Nissim, A. Smith, (2006). Calibrating noise to sensitivity in private data analysis. In: Halevi, S., Rabin, T. (eds) Theory of Cryptography. Lecture Notes in Computer Science, vol 3876. Springer, Berlin, Heidelberg. https://doi.org/10.1007/11681878_14

Eckhoff, D., and I. Wagner, (2017). Privacy in the smart city—Applications, technologies, challenges and solutions, *IEEE Commun. Surv. Tutorials*, 20, 489–516.

Elmaghraby, A. S., and M. M. Losavio, (2014). Cyber security challenges in Smart Cities: Safety, security and privacy. *J. Adv. Res.*, 5(4), 491–497. https://doi.org/10.1016/j.jare.2014.02.006

Finster S., and I. Baumgart, (2014). Privacy-aware smart metering: A survey, *EEE Commun. Surv. Tutor.*, 16(3), 1732–1745.

Gil-Garcia, J. R., (2012). Towards a smart state? Inter-agency collaboration, information integration, and beyond, *Inf. Polity*, 17(3), 269–280.

Gong, T., H. Huang, P. Li, K. Zhang, and H. Jiang, (2015). A medical healthcare system for privacy protection based on IoT, in *2015 Seventh International Symposium on Parallel Architectures, Algorithms and Programming* (PAAP). IEEE, 217–222.

Hameed, S., X. Fu, P. Hui, and N. R. Sastry, (2011). LENS: Leveraging social networking and trust to prevent spam transmission, in *Proceedings of the 19th Annual IEEE International Conference on Network Protocols*, ICNP 2011, Vancouver, BC, Canada, October 17–20, 13–18.

He, D., S. Zeadally, N. Kumar, and J.-H. Lee, (2016). Anonymous authentication for wireless body area networks with provable security, *IEEE Syst. J.*, 11(4), 2590–2601.

Hurst, W., Shone, N., El Rhalibi, A., Happe, A., Kotze, B., & Duncan, B. (2017). Advancing the micro-CI testbed for IoT cyber-security research and education. Think Mind, 129–134.

Ismagilova, E., L. Hughes, Y. K. Dwivedi, and K. R. Raman, (2019). Smart cities: Advances in research – An information systems perspective, *Int. J. Inf. Manag.*, 47, 88–100. https://doi.org/10.1016/j.ijinfomgt.2019.01.004

Jabbar, I., and S. Najim, (2016). Using fully homomorphic encryption to secure cloud computing, *Internet Things Cloud Comput.*, 4(2), 13–18.

Jing, Q., A. V. Vasilakos, J. Wan, J. Lu, and D. Qiu, (2014). Security of the internet of things: Perspectives and challenges, *Wirel. Netw.*, 20(8), 2481–2501.

Kearns, M., M. Pai, A. Roth, and J. Ullman, (2014). Mechanism design in large games: Incentives and privacy, in *Proceedings of the 5th Conference on Innovations in Theoretical Computer Science.* ACM, 403–410.

Kim, S.-H., I.-Y. Ko, and S.-H. Kim, (2017). Quality of private information (QoPI) model for effective representation and prediction of privacy controls in mobile computing, *Comput. Secur.*, 66, 1–19.

Kitchin, R., (2016). Getting smarter about smart cities: Improving data privacy and data security. Data Protection Unit, Department of the Taoiseach, Dublin, Ireland

Kolias, C., G. Kambourakis, A. Stavrou, and J. Voas, (2017). DDoS in the IoT: Mirai and other botnets, *Computer*, 50(7), 80–84.

Koop, C. E., R. Mosher, L. Kun, J. Geiling, E. Grigg, S. Long, C. Macedonia, R. C. Merrell, R. Satava, and J. M. Rosen, (2008). Future delivery of health care: Cybercare, *IEEE Eng. Med. Biol. Mag.*, 27(6), 29–38.

Koufogiannis, F., and G. J. Pappas, (2016). Diffusing private data over networks, *IEEE Trans. Control Netw. Syst.*, 5, 1027–1037.

La, Q. D., T. Q. Quek, J. Lee, S. Jin, and H. Zhu, (2016). Deceptive attack and defense game in honeypot-enabled networks for the internet of things. *IEEE Internet Things J.*, 3(6), 1025–1035.

Lee, W.-H., and R. B. Lee, (2015). Multi-sensor authentication to improve smartphone security, in *International Conference on Information Systems Security and Privacy (ICISSP)*, IEEE, 1–11.

Lee, O. J., H. L. Nguyen, J. E. Jung, T.-W. Um, and H.-W. Lee, (2017). Towards ontological approach on trust-aware ambient services, *IEEE Access*, 5, 1589–1599.

Li, N., T. Li, and S. Venkatasubramanian, (2007). t-Closeness: Privacy beyond k-anonymity and l-diversity. In: *Proceedings of the 23rd International Conference on Data Engineering (ICDE'07)*, IEEE, pp 106–115.

Li, N., D. Liu, and S. Nepal, (2017). Lightweight mutual authentication for IoT and its applications, *IEEE Trans. Sustain. Comput*, 2(4), 359–370, https://doi.org/10.1109/TSUSC.2017.2716953

Li, X., R. Lu, X. Liang, X. Shen, J. Chen, and X. Lin, (2011). Smart community: An internet of things application, *IEEE Commun. Mag.*, 49(11), 68–75.

Li, J., Y. Tao, and X. Xiao, (2008). Preservation of proximity privacy in publishing numerical sensitive data, in *Proceedings of the 2008 ACM SIGMOD International Conference on Management of Data*. ACM, 473–486.

Liu, X., K. Liu, L. Guo, X. Li, and Y. Fang, (2013). A game-theoretic approach for achieving k-anonymity in location-based services, in *INFOCOM, Proceedings IEEE*. IEEE, 2985–2993.

Luo, X., D. Zhang, L. T. Yang, J. Liu, X. Chang, and H. Ning, (2016). A kernel machine-based secure data sensing and fusion scheme in wireless sensor networks for the cyber-physical systems, *Future Gener. Comput. Syst.*, 61, 85–96.

Machanavajjhala, A., D. Kifer, J. Gehrke, and M. Venkitasubramaniam, (2007). L-diversity: Privacy beyond k-anonymity. *ACM Trans. Knowl. Discov. Data*, 1, 1.

Magno, M., T. Polonelli, L. Benini, and E. Popovici, (2015). A low cost, highly scalable wireless sensor network solution to achieve smart led light control for green buildings, *IEEE Sens. J.*, 15(5), 2963–2973.

Mahmood, Z., H. Ning, and A. Ghafoor, (2016). Lightweight two-level session key management for end user authentication in internet of things, in *2016 IEEE International Conference on Internet of Things (iThings) and IEEE Green Computing and Communications (GreenCom) and IEEE Cyber, Physical and Social Computing (CPSCom) and IEEE Smart Data (SmartData)*, IEEE, 323–327.

Meijer, A. and M. P. R. Bolívar, (2016). Governing the smart city: A review of the literature on smart urban governance, *Int. Rev. Adm. Sci.*, 82(2), 392–408.

Midi, D., A. Rullo, A. Mudgerikar, and E. Bertino, (2017). Kalis – A system for knowledge-driven adaptable intrusion detection for the internet of things, in *2017 IEEE 37th International Conference on Distributed Computing Systems (ICDCS)*, IEEE, 656–666

Mohsin, M., Z. Anwar, F. Zaman, and E. Al-Shaer, (2017). IoTChecker: A data driven framework for security analytics of internet of things configurations, *Comput. Secur.*, 70, 199–223.

Mozzaquatro, B. A., R. Jardim-Goncalves, and C. Agostinho, (2015). Towards a reference ontology for security in the internet of things, in *IEEE International Workshop on Measurements & Networking (M&N)*, IEEE, 1–6.

Mudumbe M. J., and A. M. Abu-Mahfouz, (2015). Smart water meter system for user-centric consumption measurement, in *IEEE 13th International Conference on Industrial Informatics (INDIN)*, IEEE, 993–998.

Natgunanathan, I., A. Mehmood, Y. Xiang, G. Beliakov, and J. Yearwood, (2016). Protection of privacy in biometric data, *IEEE Access*, 4, 880–892.

Nguyen, D. C., Derr, E., Backes, M., & Bugiel, S. (2019). Short text, large effect: Measuring the impact of user reviews on Android app security & privacy, in *Conference: IEEE Symposium on Security and Privacy (S&P)*. https://doi.org/10.1109/SP.2019.00012

Ning, Z., F. Xia, N. Ullah, X. Kong, and X. Hu, (2017). Vehicular social networks: Enabling smart mobility, *IEEE Commun. Mag.*, 55(5), 16–55.

Olejnik, K., I. I. Dacosta Petrocelli, J. C. Soares Machado, K. Huguenin, M. E. Khan, and J.-P. Hubaux, (2017). SmarPer: Context-aware and automatic runtime-permissions for mobile devices, in *Proceedings of the 38th IEEE Symposium on Security and Privacy* (SP), no. EPFL-CONF-226751. IEEE.

Perera, C., R. Ranjan, L. Wang, S. U. Khan, and A. Y. Zomaya, (2015). Big data privacy in the internet of things era, *IT Professional*, 17(3), 32–39. https://doi.org/10.1109/MITP.2015.34

Petrolo, R., V. Loscri, and N. Mitton, (2017). Towards a smart city based on cloud of things, a survey on the smart city vision and paradigms, *Trans. Emerg. Telecommun. Technol*, 28(1), e2931.

Quercia, D., and S. Hailes, (2010). Sybil attacks against mobile users: Friends and foes to the rescue, in *29th IEEE International Conference on Computer Communications, Joint Conference of the IEEE Computer and Communications Societies*, March 15–19, 2010, San Diego, CA (INFOCOM 2010), 336–340.

Razzaq, A., Z. Anwar, H. F. Ahmad, K. Latif, and F. Munir, (2014). Ontology for attack detection: An intelligent approach to web application security, *Comput. Secur.*, 45, 124–146.

Rezvani, M., A. Ignjatovic, E. Bertino, and S. Jha, (2015). Secure data aggregation technique for wireless sensor networks in the presence of collusion attacks, *IEEE Trans. Dependable Secure Comput.*, 12(1), 98–110.

Rodrıguez, S. S., L. Wang, J. R. Zhao, R. Mortier, and H. Haddadi, (2017). Personal model training under privacy constraints, *CoRR*, 40, 24–38. abs/1703.00380

Sehgal, A., V. Perelman, S. Kuryla, and J. Schonwalder, (2012). Management of resource constrained devices in the internet of things, *IEEE Commun. Mag.*, 50(12), 144–149.

Shamshirband, S., A. Patel, N. B. Anuar, M. L. M. Kiah, and A. Abraham, (2014). Cooperative game theoretic approach using fuzzy q-learning for detecting and preventing intrusions in wireless sensor networks, *Eng. Appl. Artif. Intell.*, 32, 228–241.

Singh, A., A. Payal, and S. Bharti, (2019). A walkthrough of the emerging IoT paradigm: Visualizing inside functionalities, key features, and open issues, *J. Netw. Comput. Appl.*, 143, 111–151, https://doi.org/10.1016/j.jnca.2019.06.013

Sun, L., J. Wang, R. Higgs, and C. Yang, (2017). Analysis on the application of dense storage technology in logistics based on the IoT, *IEEE International Conference on Computational Science and Engineering (CSE) and IEEE International Conference on Embedded and Ubiquitous Computing (EUC)*, Guangzhou, China, IEEE, pp. 214–218. https://doi.org/10.1109/CSE-EUC.2017.223

Sweeney, L., (2000). Uniqueness of simple demographics in the us population. Tech. rep., Technical report, Carnegie Mellon University.

Talpur, M. S., M. Z. A. Bhuiyan, and G. Wang, (2015). Energy-efficient healthcare monitoring with smartphones and IoT technologies. *Int. J. High Perform. Comput. Netw.*, 8(2), 186. https://doi.org/10.1504/IJHPCN.2015.070019

Tan, L., and N. Wang, (2010). Future internet: The internet of things. *3rd International Conference on Advanced Computer Theory and Engineering(ICACTE)*, Chengdu, China, IEEE, pp. V5-376–V5-380. https://doi.org/10.1109/ICACTE.2010.5579543

Tao, M., J. Zuo, Z. Liu, A. Castiglione, and F. Palmieri, (2016). Multi-layer cloud architectural model and ontology-based security service framework for IoT based smart homes, *Future Gener. Comput. Syst.*, 78(3), 1040–1051.

Tsai, C.-W., C.-F. Lai, M.-C. Chiang, L. T. Yang et al., (2014). Data mining for internet of things: A survey. *IEEE Commun. Surv. Tutor.*, 16(1), 77–97.

van Zoonen, L., (2016). Privacy concerns in smart cities, *Gov. Inf. Q.*, 33(3), 472–480, https://doi.org/10.1016/j.giq.2016.06.004

Walravens, N., (2012). Mobile business and the smart city: Developing a business model framework to include public design parameters for mobile city services, *J. Theor. Appl. Electron. Commer. Res.*, 7(3), 121–135.

Wang, K., M. Du, S. Maharjan, and Y. Sun, (2017). Strategic honeypot game model for distributed denial of service attacks in the smart grid, *IEEE Trans Smart Grid*, 8(5), 2474–2482. https://doi.org/10.1109/TSG.2017.2670144

Wang, Y., J. Wan, J. Guo, Y.-M. Cheung, and P. C. Yuen, (2017b). Inference-based similarity search in randomised Montgomery domains for privacy preserving biometric identification, *IEEE Trans. Pattern Anal. Mach. Intell.*, 40, 1611–1624.

Wang, K., Y. Xu, A. W.-C. Fu, and R. C.-W. F Wong, (2009). f-anonymity: When quasi-identifiers are missing, in *IEEE 25th International Conference on Data Engineering, 2009. ICDE'09,* IEEE, 1136–1139.

Wijesekera, P., A. Baokar, L. Tsai, J. Reardon, S. Egelman, D. Wagner, and K. Beznosov, (2017). The feasibility of dynamically granted permissions: aligning mobile privacy with user preferences, in 2017 *IEEE Symposium on Security and Privacy* (SP). IEEE, 1077–1093.

Wong, R. C.-W., A. W.-C. Fu, K. Wang, and J. Pei, (2007). Minimality attack in privacy preserving data publishing, in *Proceedings of the 33rd International Conference on Very Large Data Bases, VLDB Endowment*, pp. 543–554.

Wu, J., K. Ota, M. Dong, J. Li, and H. Wang, (2016). Big data analysis-based security situational awareness for smart grid, *IEEE Trans. Big Data.*, 4(3), 408–417.

Xiao, L., Y. Li, G. Han, G. Liu, and W. Zhuang, (2016). Phy-layer spoofing detection with reinforcement learning in wireless networks, *IEEE Trans. Veh. Technol.*, 65(12), 10037–10047.

Xing, K., C. Hu, J. Yu, X. Cheng, and F. Zhang, (2017). Mutual privacy preserving k-means clustering in social participatory sensing, *IEEE Trans. Ind. Inf.*, 13(14), 2066–2076.

Xu, G., Y. Cao, Y. Ren, X. Li, and Z. Feng, (2017). Network security situation awareness based on semantic ontology and user-defined rules for internet of things, *IEEE Access*, 5(21), 046–21056.

Xynos, K., I. Sutherland, and A. Blyth, (2013). *Effectiveness of Blocking Evasions in Intrusion prevention System.* University of South Wales, 1–6.

Yassine, A., A. A. N. Shirehjini, and S. Shirmohammadi, (2015). Smart meters big data: Game theoretic model for fair data sharing in deregulated smart grids, *IEEE Access*, 3, 2743–2754.

Yu, S., (2016). Big privacy: Challenges and opportunities of privacy study in the age of Big Data, *IEEE Access*, 4, 2751–2763.

Yu, S., G. Gu, A. Barnawi, S. Guo, and I. Stojmenovic, (2015). Malware propagation in large-scale networks, *IEEE Trans. Knowl. Data Eng.*, 27(1), 170–179.

Yu, R., J. Kang, X. Huang, S. Xie, Y. Zhang, and S. Gjessing, (2016). MixGroup: accumulative pseudonym exchanging for location privacy enhancement in vehicular social networks, *IEEE Trans. Dependable Secure Comput.*, 13(1), 93–105.

Yue, X., H. Wang, D. Jin, M. Li, and W. Jiang, (2016). Healthcare data gateways: found healthcare intelligence on blockchain with novel privacy risk control, *J. Med. Syst.*, 40(10), 218.

Zhang, Z.-K., M. C. Y. Cho, and S. Shieh, (2015). Emerging security threats and countermeasures in IOT, in *Proceedings of the 10th ACM Symposium on Information, Computer and Communications Security.* ACM, 1–6.

Zhang, K., J. Ni, K. Yang, X. Liang, J. Ren, and X. S. Shen, (2017). Security and privacy in smart city applications: Challenges and solutions, *IEEE Commun. Mag.*, 55(1), 122–129.

Zhou, B., and J. Pei, (2008). Preserving privacy in social networks against neighborhood attacks, in *IEEE 24th International Conference on Data Engineering, (2008). ICDE 2008,* IEEE, 506–515.

Zhou, L., C. Su, W. Chiu, and K.-H. Yeh, (2017). You think, therefore you are: Transparent authentication system with brainwave-oriented bio-features for IOT networks, *IEEE Trans. Emerg. Top. Comput.*, 8(2), 303–312.

Index

M

N

O

P

R

S

T

U

W

Z